酒类英语

主　编：云虹　吕军录

副主编：梁勇　蒋华应　黄燕　蔡玲凌

编　者：许静　王庆　王智国　徐乾朗　姚楠　阮琪

外语教学与研究出版社
FOREIGN LANGUAGE TEACHING AND RESEARCH PRESS
北京 BEIJING

图书在版编目(CIP)数据

酒类英语 / 云虹等主编；许静等编. -- 北京：外语教学与研究出版社，2019.8
（2021.8 重印）
 ISBN 978-7-5213-1140-2

Ⅰ. ①酒… Ⅱ. ①云… ②许… Ⅲ. ①酒－英语 Ⅳ. ①TS262

中国版本图书馆 CIP 数据核字 (2019) 第 185979 号

出 版 人　徐建忠
责任编辑　杨芳莉
责任校对　黄　骥
封面设计　郭　莹
版式设计　付玉梅
出版发行　外语教学与研究出版社
社　　址　北京市西三环北路 19 号（100089）
网　　址　http://www.fltrp.com
印　　刷　三河市北燕印装有限公司
开　　本　787×1092　1/16
印　　张　11.5
版　　次　2020 年 3 月第 1 版　2021 年 8 月第 2 次印刷
书　　号　ISBN 978-7-5213-1140-2
定　　价　45.90 元

购书咨询：(010) 88819926　电子邮箱：club@fltrp.com
外研书店：https://waiyants.tmall.com
凡印刷、装订质量问题，请联系我社印制部
联系电话：(010) 61207896　电子邮箱：zhijian@fltrp.com
凡侵权、盗版书籍线索，请联系我社法律事务部
举报电话：(010) 88817519　电子邮箱：banquan@fltrp.com
物料号：311400001

四川省社科规划项目成果（SC18WY024）
五粮液酒文化国际传播项目成果（HX2019209）

中国白酒文化传播研究丛书
（四川省宜宾五粮液集团股份有限公司资助）

编委会成员名单

编委会主任　总主编
云　虹

编委会委员
王洪渊
廖国强
吕军录
黄　燕
田学梅

前 言

中国酒文化历史悠久,源远流长,内涵丰富,博大精深,酒在人们日常交往中常常发挥着重要的作用。俗语有言:"无酒不成席",无论是节日庆典,家人亲朋聚会,还是国家领导人会面,常常都离不开酒的参与。古今中外,酒渗透到了社会生活的方方面面,因此,酒不仅是一种饮品,更是一种文化象征,在人类历史长河中占据了重要地位。伴随着人类文明的发展和进步,酒文化应运而生。酒文化是中国文化百花园中的重要组成部分,它枝繁叶茂,芳香四溢,至今已延续了数千年,每个阶段的发展,都带有鲜明而深刻的时代烙印。千百年来,文人墨客饮酒作诗,借酒明志,留下了无数的佳作。酒文化里也体现着无数英雄豪杰的英勇壮举和劳动人民朴素的日常生活。同时,酒也和音乐、诗词、书法、美术、影视相融相兴,共同成就了中国璀璨的酒文化。酒文化作为中国文化的重要组成部分,我们不仅要重视其在国内的传承,更应该将其推向世界。一方面,随着我国国际影响力的不断加强,中国文化输出的不断加深,中国的酒和酒文化在世界上赢得了越来越多的关注。另一方面,秉承"民族的即是世界的"理念,将中国的酒与酒文化介绍给全世界,进行跨文化传播与交流也势在必行。

本书介绍了酒的起源与发展、分布与酿造、酒品的分类与品鉴、名酒的典故与鉴赏,以及科学、健康饮酒新概念等,还包括了中国酒礼、酒俗等内容,有助于学习者全面而详尽地了解中国酒文化知识,加深对酒文化的理解,提高文化素养。学习者通过学习使用本教材能够掌握酒类英语相关词汇,能用英语进行相关专业以及酒文化方面的基本交流。本书包含八个单元,每单元提供一篇精读文章和一篇泛读文章。本书可作为大学外语分级教学后续课程教材,也可作为学习酒文化的教材,供对酒文化感兴趣的读者使用。

四川省宜宾五粮液集团股份有限公司资助了本书出版,在此表示感谢。

Contents

Unit 1 Overview of Liquor-making in China and in the West p1	Preview p1	**Section A** The Brief History of Liquor-making in China p2	**Section B** The Origin of Liquor-making in the West p8
Unit 2 Legends and Culture of Liquor in China p15	Preview p15	**Section A** Legends of the Origin of Liquor in China p16	**Section B** The Culture of Wuliangye and the Culture of Moutai p22
Unit 3 Drinking and Health p27	Preview p27	**Section A** Benefits of Drinking p28	**Section B** How to Drink in a Healthy Way p36
Unit 4 Baijiu Map p41	Preview p41	**Section A** Main Distribution of Baijiu and the Most Representative Brands of the Regions p42	**Section B** China's *Bordeaux*: the Golden Triangle of Baijiu on the Upper Reach of the Yangtze River p49
Unit 5 Manufacture and Classification of Baijiu p57	Preview p57	**Section A** Manufacture of Baijiu p58	**Section B** Classification of Baijiu p64
Unit 6 Tasting p71	Preview p71	**Section A** Enjoying Tasting Baijiu p72	**Section B** Aroma, the Soul of Baijiu p83
Unit 7 Drinking Etiquette p89	Preview p89	**Section A** Drinking Etiquette p90	**Section B** Drinking Etiquette of Ethnic Minorities p97
Unit 8 The Selection of Famous Baijiu p103	Preview p103	**Section A** The Selection of Famous Baijiu p104	**Section B** The Minor Styles of Baijiu p117

UNIT 1 Overview of Liquor-making in China and in the West

Preview

Liquor is a changeable spirits. She is as hot as fire, while sometimes she is as cold as ice; she is as soft as *damask*, while now and then she is as sharp as steel blade. She can make one open-minded and unleash one's potential. She can make one forget the sorrows and worries temporarily and enjoy the absolute happiness and freedom in the drunken world. And she can make one reveal the true nature and tell the truth. In the long history of human culture, liquor is not only a kind of drink, but also has become the symbol of human culture.

Section A

The Brief History of Liquor-making in China

China is one of the countries which first made liquor in the world. At the same time, she is also the cradle of spirits culture. Liquor-making enjoys a very long history in China. In the thousands of years of history of Chinese civilization, the development of liquor-making and Chinese culture has become basically *synchronized*. In China, the beginning of liquor-making can be traced back to the prehistoric period. The original alcohol was a kind of organic material from sugar-containing substance naturally formed by *yeast*. There are plenty of sugar-containing wild fruits in nature, and yeasts are attached to the air, dust and peel. Under the proper moisture and temperature, yeast can turn flesh into *pulp* and form wine naturally. The history of wine-making in China began about 40,000-50,000 years ago in the *Paleolithic Period*. At that time, people had enough food to support their basic needs, so they were able to imitate the biological wine-making process of nature. The earliest wine-making activity was simply a mechanical repetition of nature's *brewing* process. The truly purposeful liquor-making actually began after mankind entered the *Neolithic Age* with the emergence of agriculture. During this period, human beings had sufficient grains and the finely-made pottery utensils, which made liquor-making possible. During the period of Chinese Yangshao Culture (5,000 B.C.-3,000 B.C.), farming tools, namely agriculture, appeared in China, which made it possible for people to make liquor by *cereals*. According to the *Outline of Chinese History*, the period of Chinese Yangshao Culture was the "budding" period during which ancient Chinese made liquor by cereals. At that time, our ancestors made liquor with *distiller's* yeast (*germinated* grain). Among the earthenware unearthed from the site of Chinese Longshan Culture (2,500 B.C.–2,000 B.C.), many of them, such as *Zun* (尊), *He* (盉), goblets, small wine pots and so on, reflected that liquor-making was prevailing during that period. Alcohol is created by people of all nationalities world around in the long history. The oldest wine in the world is the wine unearthed in *Samaria,* Iran. Although it was made more than 3,000 years ago, it is still *aromatic.* The oldest liquor found in China is the imperial liquor of the Han Dynasty unearthed in Xi'an. According to textual analysis by experts, it is made from grain and is still *mellow* and drinkable. The Chinese character "酒" has long appeared in Chinese oracle-bone inscriptions and so do other characters, such as *Li* (醴), *Zun* (尊), *You* (酉) and so on, which tells us a lot about the long history of liquor-making in China. There are numerous lines about liquor and liquor-making in historic books and literary works. For example, there is such a line in *The Book of Songs* as "I am drunk with your liquor and enlightened by your virtue". And there are more lines about the liquor-drinking custom, for instance, there is such a line in *The Book of Rites* as "Liquor can prolong one's life", and "Liquor is used to show etiquette and you cannot drink too much" in *Zuo Zhuan.* All of these show that liquor had many uses.

    Chinese liquor-making has a long history with great varieties, and various products enjoy a good reputation both at home and abroad. The earliest liquor in China were made from fruit or rice. From the Xia, Shang, Zhou, Qin and Han Dynasties to the Tang and Song Dynasties, people made liquor from fruits and grains, which were first steamed or cooked and then mixed with distiller's yeast, later stored the mixture for a period of time and finally got the liquor through *milling-diffusion.* With the progress of human society, the liquor-making technology has also been further improved. Chinese people have invented *distillation method* in the Song Dynasty. Since then *Chinese spirits* have become the main liquor in Chinese people's daily life. Thanks to the further development of liquor-making technology, there are great varieties of liquor products, such as Chinese spirits, beer, wine, yellow wine (rice wine), medicinal liquor and so on. There are ten most famous brands among the Chinese spirits category, which are Moutai, Wuliangye, Jiannanchun, Luzhou Laojiao Tequ, Xifengjiu, Fenjiu, Gujinggong Liquor, Dongjiu, Yanghe Daqu and Langjiu.

For thousands of years, in the *vicissitudes* of history, the creative Chinese people have made many famous alcohol products with local characteristics and customs. In addition, different nationalities in different regions have developed various drinking customs and rituals. All of these have made China a country with rich and famous liquor. Liquor has *permeated* the 5,000 years of Chinese civilization. It has been playing a very important role in Chinese people's daily life, from literary and artistic creation to cultural recreation, and from eating and cooking to health care.

New Words and Expressions

1 synchronize [ˈsɪŋkrəˌnaɪz] *vi.* 同时发生
2 yeast [ji:st] *n.* 酵母；发酵剂；酒曲
3 pulp [pʌlp] *n.* 果浆
4 Paleolithic Period 旧石器时代
5 Neolithic Age 新石器时代
6 brew [bru:] *v.* 酿造；酝酿；策划

7 cereal [ˈsɪəriəl] *n.* 谷物

8 distiller [dɪˈstɪlə] *n.* 蒸馏器

9 germinate [ˈdʒɜːmɪˌneɪt] *v.* 发芽；开始生长

10 aromatic [ˌærəˈmætɪk] *adj.* 芳香的

11 mellow [ˈmeləʊ] *adj.* 成熟的；醇的

12 vicissitudes [vaɪˈsɪsəˌtjuːdz] *n.* 变化；变迁；兴败；盛衰

13 permeate [ˈpɜːmɪˌeɪt] *v.* 弥漫；渗透

14 distillation [ˌdɪstɪˈleɪʃn] *n.* 蒸馏

15 spirits culture 酒文化

16 milling-diffusion 压榨渗透法

17 distillation method 蒸馏法

18 Chinese spirits 中国白酒

Proper Names

1 Chinese Yangshao Culture 中国仰韶文化，是黄河中游地区一种重要的新石器时代彩陶文化，其持续时间大约在公元前5000年至公元前3000年，分布在整个黄河中游，在今天的甘肃省到河南省之间。

2 Chinese Longshan Culture 中国龙山文化，泛指中国黄河中下游地区约新石器时代晚期的一类文化遗存，其持续时间大约在公元前2500年至公元前2000年，分布于黄河中下游的河南、山东、山西、陕西等省。

3 Samaria 撒玛利亚（伊朗）

4 *Outline of Chinese History* 《中国史稿》

5 *The Book of Songs* 《诗经》

6 *The Book of Rites* 《礼记》

7 *Zuo Zhuan* 《左传》

Exercises

I. Answer the following questions.

1 Why is China the cradle of spirits culture?

2 What is the relationship between liquor-making and Chinese culture?

3 When did Chinese people begin making liquor?
4 How did Chinese people know the method of liquor-making?
5 When did the purposeful liquor-making begin in China?
6 Where did people find the oldest wine in the world? How about the oldest liquor found in China?
7 Are there any lines about liquor and liquor-making in Chinese historic books or literary works? Give some examples.
8 What was the earliest wine made from in China?
9 When did Chinese people invent the distillation method?
10 List the top ten famous brands among the Chinese spirits category.

II. Discuss the following questions.

1 What do you think of the role of China in the liquor-making history of the world?
2 Of the top ten famous brands of Chinese spirits, which do you like best and why?

III. Match the items in column A with those in column B.

A	B
1. 合适的湿度和温度	____a. pulp
2. 酒曲	____b. Chinese spirits
3. 酒文化	____c. medicinal liquor
4. 蒸馏法	____d. imperial liquor
5. 甲骨文	____e. distillation method
6. 压榨渗透法	____f. milling-diffusion
7. 果浆	____g. oracle-bone inscription
8. 中国白酒	____h. proper moisture and temperature
9. 御酒	____i. spirits culture
10. 药酒	____j. distiller's yeast

IV. Fill in the following blanks with the words given below. Change the form of the words when necessary.

| permeate | enlighten | germinate | unscrupulous | vicissitude |
| aromatic | brew | ubiquitous | prolong | prevail |

1. My brother thought the world was flat until I _____ him.
2. We are gradually beginning to realize that the potential for life is _____ throughout our galaxy.
3. Every year this time, the misty weather _____ in this part of the country.
4. The smell of her perfume _____ the whole room.
5. An idea for a novel began to _____ in her mind.
6. This TV program reflects the _____ of the society.
7. The meeting may be _____ into the evening because so many problems have to be solved tonight.
8. He was utterly _____ in his competition with rival firms.
9. In some remote places, people usually _____ their own beer at home.
10. In order to make delicious dishes, my mother now and then uses _____ herbs in cooking.

V. Translate the following passage into English.

黄酒是中国特产，也被称为米酒（rice wine），是世界三大酿造酒（黄酒、葡萄酒和啤酒）之一。黄酒酿造技术是东方酿造的典型代表。绍兴黄酒是历史最悠久、最有代表性的产品。它是一种以稻米为原料酿制成的粮食酒。不同于白酒，黄酒没有经过蒸馏，酒精含量低于20%。不同种类的黄酒颜色亦不同，有的是米色，有的是黄褐色，有的是红棕色。山东即墨老酒是北方粟米黄酒的典型代表；福建龙岩沉缸酒、福建老酒是红曲黄酒的典型代表。

VI. Write a passage according to the given outline in no less than 150 words.

Introduce one of the top ten brands of Chinese spirits to your American friend. Your introduction should include the following parts:

1. The brief history of the brand;
2. The characteristics of the brand;
3. Its reputation at home and abroad.

Section B

The Origin of Liquor-making in the West

There is no specific historical record about liquor-making in the West. Most historians believe that liquor-making in the West started from making wine, which appeared firstly in Persia and later on the Greek island of Crete and then it was spread to France and elsewhere through wars and trades. So liquor-making in the West can be traced back to the ancient Persia (now Iran).

By about 3,000 B.C., the Egyptians had already started making wine. A large number of precious cultural *relics* (especially *reliefs*) found in ancient Egyptian tombs clearly *depicted* the cultivating and harvesting of grapes, and making of wine by the ancient Egyptians. The earliest cultivation of grapes began about 7,000 years ago in today's South Caucasus, Central Asia, Syria and Iraq. Later, with wars and immigrants, the techniques were spread to Egypt firstly, then Greece and other countries and regions. But the real available information comes from the tremendous relics found in Egyptian tombs. In the *excavated* tombs in Nile Valley area, *archaeologists* found a kind of *clay* pots with small bottom, thick *belly*, and large upper neck, which, by textual research, were earthen pots used by the ancient Egyptians to hold wine or oil. The reliefs, which enjoy a long history, have vividly described the cultivation and harvest of grapes, the winemaking process and the drinking scenes. In addition, the wine bottles produced at the time of ancient Egypt were also *engraved* with the word *Ilp* (meaning "wine" in Egyptian). Hugh Johnson, a famous writer who focuses on the subject of wine once wrote, "There were excellent experts on wine in Ancient Egypt, who, just like the producers and *dealers* of Sherry or the agents of Bordeaux wine in the 20th century, could confidently and professionally identify the quality of wine."

Wine has its origin with human civilization. There are stories about wine in ancient *myths* and legends of the world, and there are traces of the development of human history in the history of wine.

Starting with grapes, people are constantly trying to make liquor with a variety of *ingredients*. There are thousands of kinds of liquor around the world, and the *raw* materials and *alcohol content* of wine also vary greatly. Therefore, people have tried different ways to classify them so as to make it easier to understand and remember. If classified by raw materials, it can be roughly divided into seven categories: grain-based liquor, spice-herb-flavored liqueurs, fruit wine, milk and egg wine, plant wine, honey wine and mixed wine. According to serving time, there are pre-meal wine (or appetizer), table wine, after-meal wine and special drink; according to alcohol content, there are mild wine, medium wine and strong wine. However, if classified by production process it can be divided into three categories: 1. *Fermented* liquor. Wine, beer, rice wine, *cider*, etc. are included. 2. Distilled liquor. Whisky, brandy, vodka, rum, tequila and Chinese spirits belong to this category.

3. *Refined* and *synthesized* liquor. In this group, there are gin, liqueur, vermouth, bitters, Green Bamboo Leaf Liquor, Ginseng Liquor and so on. Among all of these methods of classification, the most popular and widely used one is sorting liquor by its production technology.

There are many world-famous liquor brands, such as the Remy Louis XIII (France), which symbolizes wealth and status; Bacardi Rum (West Indies), which is honored as the source of excitement and *uproar*; vodka (Russia), the burning water of life; Chivas Regal (Scotland), the royal drink; Hennessy (France), the brandy *mythology*; Martell (France), the flowing gold; Jack Daniel's (U.S.A.), the *tranquil* feelings of countryside; and Tequila (Mexico), the blazing fire. From the colorful and distinctive names, it is not difficult to see that liquor plays a significant role in the daily life and interpersonal communication of Westerners.

New Words and Expressions

1. relic [ˈrelɪk] *n.* 遗迹
2. relief [rɪˈliːf] *n.* 浮雕
3. depict [dɪˈpɪkt] *vt.* 描述；描绘
4. excavate [ˈekskəˌveɪt] *v.* 发掘（古物）
5. archaeologist [ˌɑːkiˈɒlədʒɪst] *n.* 考古学家
6. clay [kleɪ] *n.* 黏土；泥土
7. belly [ˈbeli] *n.* 腹部
8. engrave [ɪnˈɡreɪv] *v.* 雕刻
9. dealer [ˈdiːlə] *n.* 商人；经销商
10. myth [mɪθ] *n.* 神话
11. ingredient [ɪnˈɡriːdiənt] *n.* 成分；原料；配料

12 raw [rɔː] *adj.* 未加工的；生的
13 alcohol content 酒精含量
14 ferment [ˈfɜːment] *v.* (使) 发酵
15 cider [ˈsaɪdə] *n.* 苹果酒
16 refine [rɪˈfaɪn] *v.* 精炼；改进
17 synthesize [ˈsɪnθəˌsaɪz] *v.* 合成；综合
18 ginseng [ˈdʒɪnˌseŋ] *n.* 人参；高丽参
19 uproar [ˈʌpˌrɔː] *n.* 骚动；喧嚣
20 mythology [mɪˈθɒlədʒi] *n.* 神话
21 tranquil [ˈtræŋkwɪl] *adj.* 安静的；宁静的

Proper Names

1 Persia [ˈpɜːʃə] *n.* 波斯（现在的伊朗）
2 Hugh Johnson [hjuː ˈdʒɔnsn] 休·约翰逊（1939-）英国著名葡萄酒评论家，被公认为当今世界首屈一指的葡萄酒史权威和最畅销的酒类指南作家。代表著作：《葡萄酒》(*Wine*)，《世界葡萄酒地图》(*World Atlas of Wine*)，《葡萄酒袖珍书》(*Pocket Wine Book*)，《葡萄酒指南》(*Wine Companion*)，《葡萄酒的故事》(*The Story of Wine*)
3 Bordeaux [bɔːˈdəʊ] *n.* 波尔多葡萄酒
4 vodka [ˈvɒdkə] *n.* 伏特加酒
5 rum [rʌm] *n.* 朗姆酒
6 tequila [tɪˈkiːlə] *n.* 龙舌兰酒
7 gin [dʒɪn] *n.* 杜松子酒
8 liqueur [lɪˈkjʊə] *n.* 利口酒，烈性甜酒
9 vermouth [ˈvɜːməθ] *n.* 苦艾酒（用作开胃酒），味美思酒
10 Remy Louis XIII [remi luisˌθɜːˈtiːn] 人头马路易十三
11 Bacardi Rum [bəkɑːdi rʌm] 百家得朗姆酒
12 Chivas Regal [ˈtʃvas] 芝华士酒
13 Hennessy [ˈhenisi] *n.* 轩尼诗酒
14 Martell [ˌmɑː(r)ˈtel] 马爹利酒
15 Jack Danny [ˈdʒæk ˈdæni] 杰克丹尼酒

Exercises

I. Choose the best answer to each of the following questions.

1. According to most historians, when and where did liquor-making begin?
   A) 6,000 B.C.; Greek.
   B) 3,000 B.C.; France.
   C) 7,000 B.C.; Iran.
   D) 6,000 B.C.; Persia.

2. When did the earliest cultivation of grapes begin?
   A) 6,000 years ago.
   B) 3,000 years ago.
   C) 7,000 years ago.
   D) 5,000 years ago.

3. What did the archaeologists find in the excavated tombs in Nile Valley area?
   A) A kind of earthen pots used for milk or oil.
   B) A kind of ceramic pots used for wine or oil.
   C) A kind of clay pots used for wine or oil.
   D) A kind of glass pots used for wine or tea.

4. What is the most convincing evidence that tells the ancient Egyptians began to make wine in about 3,000 B.C.?
   A) The earthen pots found in ancient Egyptian tombs.
   B) The wall painting found in ancient Egyptian tombs.
   C) The wine bottles found in ancient Egyptian tombs.
   D) The reliefs found in ancient Egyptian tombs.

5. What does the word "Ilp" represent?
   A) "Wine" in French.
   B) "Water" in Egyptian.
   C) "Wine" in Egyptian.
   D) "Whisky" in Persian.

6. Which of the following is classified by raw materials?
   A) Table wine.
   B) Grain-based liquor.
   C) Mild wine.
   D) Ginseng liquor.

7 Which classification method is the most popular and widely used one?
   A) By drinking time.   B) By raw materials.
   C) By alcohol content.   D) By production process.

8 Which of the following is regarded as the symbol of wealth and status?
   A) Bacardi Rum.   B) The Remy Louis XIII.
   C) Chivas.   D) Jack Danny.

II. Translate the following passage into English.

在西方，酒的产生有一个有趣的传说。据传，有一位古波斯国王，把吃不完的葡萄藏在瓶中密封起来，并写上"毒药"两字，以防他人偷吃。国王日理万机，很快便把这件收藏给忘记了。后来，有位失宠的妃子凑巧看到这瓶"毒药"，便产生轻生之念。打开后，里面颜色古怪的液体也很像毒药，她就喝了几口。在等死的时候发觉不但不痛苦，反而有种舒适陶醉的感觉。于是她将这事呈报国王。国王大为惊奇，一试果不其然。这就是葡萄酒的来历。

UNIT 2
Legends and Culture of Liquor in China

Preview

China is the hometown of Baijiu as well as the origin of Baijiu culture, particularly the earliest country to enjoy liquor-brewing technology in the world. In the course of history, the improvement of liquor-brewing technology in China has always been accompanied by the enrichment of the liquor culture. Baijiu occupied an important position in the course of Chinese history, and has become an integral part of the Chinese history and culture.

Section A

Legends of the Origin of Liquor in China

Liquor enjoys a long history, which could *be traced back* to ancient times. Liquor has been playing a significant role in the development of Chinese society. Moreover, Liquor is *indispensable* for important *banquet*s, social events and daily events, and has comprised an *integral* part of the Chinese history and the Chinese culture.

There is no fixed date about the origin of Liquor; indeed, the history of liquor is longer than that of the recorded history of states and dynasties. It has been said that "liquor came into existence before states". There are numerous legend on the origin of liquor, which could be mainly divided into three versions:

1 The Heaven-created Version

The version of liquor created by the Heaven is mainly from the ancient folk legend, *documentary records* and poetry. The story given in the book *Map of Liquor* about the origin of liquor was rather mythical, saying that "Chinese liquor was the *masterpiece* of the liquor star in heaven." As a result, Chinese people in ancient times believed in that "The liquor star *took charge of* liquor". It has been proven once again in the poetry of the men of letters. Li Bai known as "the Immortal Poet" once wrote in his poem *Drinking Alone in the Moon*: "If the heaven doesn't love liquor, the liquor star couldn't be in heaven". There is another saying in *"Letter to Cao Cao on Prohibition of Liquor"*: "The glory of the liquor star is beloved by the heaven, the place of liquor is listed as a capital." From the ancient times, People confirmed this version "The liquor star created liquor". However, this kind of version *is lacking in* the scientific evidence as well as the *verified* argument. To be frank, this is a literally *exaggerated* saying.

2 The Ape-created Version

According to some ancient *fossil* records, apes not only love to drink liquor, but also like to make it. There are countless records in Chinese classics. Li Tiaoyuan, a famous scholar in the Qing Dynasty, once wrote in his book "There were lots of apes so people could get liquor from stone *crannies*. Apes added some grains with kinds of flowers to brew liquor, which was *pungent* and difficult to be got." Li baihua, another famous scholar in the Ming Dynasty, also wrote similar words in his works, "there were many apes, which picked up fruits in spring and summer in order to brew into liquor after storing them in the hole of stone for a period of time". Different people have recorded the similar content to verify the fact, that is, in the settlement of apes, it is easy to find something like Liquor. Apes *discarded* or stored fruits in stone crannies or tree holes, where the sugar content was naturally *fermented* into juice. However, such statements do not bear any cultural significance in the strict sense of the term.

3 The Legend of Yidi and Dukang

It was said that Yidi, a very beautiful woman during Dayu's times, created liquor. In the second century B.C., *Mister Lü's Spring and Autumn Annals* stated, "Yidi brewed liquor". It was an interesting legend: When Yidi cooked, she suddenly smelt the aroma of cereal. Thus, she fermented the cereals and sent the juice to King Dayu to try. The King found it very sweet and *aromatic*. After continuous drinking, he developed an addiction and gradually neglected his duties. Later when he realized that drinking the juice could disturb his work, he decided to stay away from the juice and

Lady Yidi. Meanwhile, he predicted that some kingdoms would be ruined in the future because of liquor. This story is consistent with what had been recorded in *Strategies of the Warring States* edited by Liu Ce in the Han Dynasty. History books mentioned Yidi many times in order to stress the fact that Yidi was the ancestor of liquor-making. Here was another saying about Dukang who was also regarded as the ancestor of liquor. There were several versions about the identity of Dukang. One version held that Dukang was Shaokang, who was the sixth king of the Xia Dynasty. Another Version argued that Dukang lived during the Zhou Dynasty. The former version was likely to be closer to the truth on the basis of the historic records. It was said that Dukang would put the leftover in the tree hollows in his *mulberry grove*. After a short period, the fermented food emitted aroma. This was the original method of making liquor. Because of this, people treated Dukang as the founding father of liquor and his name "Dukang" became synonymous to liquor.

To sum up, in the thousands of years long history, liquor has developed in stages. The first stage is from 4,000 B.C. to 2,000 B.C., which covers from the early days of Yangshao Culture in *the Neolithic Age* to the early years of the Xia Dynasty. This period was the late stage of primitive society, in which ancient people treated liquor as a drink with a magic element. This stage lasted almost 2,000 years and was the enlightenment period for traditional Baijiu. Brewing water with fermented grains was the main form of making liquor in that period. From the Xia Dynasty to the Qin Dynasty, it was the growth period for the traditional Chinese Baijiu. At that time, the liquor-making industry has been greatly developed and paid close attention by the government. The government had set up the liquor-making institution and controlled the liquor industry. The third stage was from the Qin Dynasty to the Northern Song

Dynasty. This stage witnessed the maturation of the traditional Baijiu. Drinking was not only prevalent in the upper class, but also popular among ordinary families. In this stage, the prosperity of the Han and Tang Dynasties as well as the rise of trade among European, Asian and African countries spurred the infiltration in cultures between the East and the West, which laid the foundation for the invention and development of Chinese Baijiu. The last stage was from the Northern Song Dynasty to the late Qing Dynasty. In the meantime, the distiller in Western regions was introduced to China, which contributed to the invention of Baijiu, so the stage featured the improvement period of liquor. From then on, Baijiu has found a way into every aspect of people's life and become a widespread drink, with constant innovation and development to meet people's ever-growing needs.

New Words and Expressions

1. be traced back to 追溯到
2. indispensable [ˌɪndɪˈspensəbl] adj. 不可缺少的
3. banquet [ˈbæŋkwɪt] n. 宴会
4. integral [ˈɪntɪɡrəl] adj. 完整的
5. documentary records 文献记载
6. *Map of Liquor*《酒谱》
7. masterpiece [ˈmɑːstəpiːs] n. 杰作
8. take charge of 负责管理
9. Drinking Alone in the Moon《月下独酌其二》（唐代李白的诗歌）
10. Letter to Cao Cao on Prohibition of Liquor《与曹操论禁酒书》（三国孔融的散文）
11. be lacking in 缺少

12　exaggerate [ɪgˈzædʒəˈreɪt] *v.* 夸张
13　fossil [ˈfɒsl] *n.* 化石
14　cranny [ˈkrænɪ] *n.* 缝隙
15　pungent [ˈpʌndʒənt] *adj.* 辛辣
16　discard [dɪsˈkɑːd] *v.* 丢弃
17　mulberry grove 桑树林
18　bear... significance in sth. 具有什么作用

Exercises

I. Answer the following questions.

1　How did apes create wine?
2　Why did the King decide to be away from Yidi?
3　Why did traditional Baijiu gain fast development in the third stage?
4　What are the functions of Baijiu in Chinese society?
5　What are poets' attitudes to wine?
6　In the first stage, what happened to the development of traditional Baijiu?
7　In the fourth stage, what has been introduced into China?

II. Discuss the following questions.

1　Which version about liquor-making do you prefer?
2　Can you briefly introduce the four stages of the development of traditional Baijiu?

III. Match the items in column A with those in column B.

A	B
1. flavor	____a. 主体香
2. rich	____b. 落口
3. main aroma	____c. 酒花
4. distillate bubble	____d. 酱香型
5. residual taste	____e. 辣味
6. laobaigan-flavor style	____f. 浓香型
7. piquancy	____g. 风味
8. swallow	____h. 老白干
9. jiang-flavor style	____i. 浓厚
10. strong-flavor style	____j. 余味

IV. Fill in the following blanks with the words given below. Change the form of the words when necessary.

take charge of	be lacking in	integral	be traced back
discard	verified	indispensable	banquet
bear... significance in sth		exaggerate	

1. Dr. Smith is assigned to _____ the department.
2. Many areas in our country _____ natural resources.
3. Taiwan is a (an) _____ part of China.
4. The history of wine can _____ to millenniums (千年) ago.
5. It is not _____ to say that China is rich in minerals.
6. Tonight there will be a (an) _____ in the playground.
7. Humans are _____ to water.
8. Don't _____ the unused papers anywhere.
9. Nobody could _____ his words in the court so he was arrested.
10. This poem _____ in the development of Chinese history.

V. Translate the following passage into English.

King Yao was one of the five kings in ancient times. As the Chinese legend had it, Yao was the reincarnation of a living dragon and he was very sensitive to nimbus. Attracted by the nimbus of the Water-dropping Pond, he led his people to settle down near the pond and engaged in farming activities and led a peaceful life. In order to express their gratitude to Heaven, King Yao selected the best cereals and soaked them in clear pond water, removed the impurities from the essence and mixed the water together to make offering water. This clean, pure and aromatic water was the earliest form of Baijiu.

VI. Write a passage according to the given outline in no less than 150 words.

On the History of Chinese Baijiu

1) 关于白酒的起源的基本介绍；
2) 详细介绍你认为最科学的一种；
3) 举例说明（可借鉴某种白酒的历史，比如五粮液）。

Section B

The Culture of Wuliangye and the Culture of Moutai

Wuliangye and Moutai enjoy high status and popularity among Baijiu brands. Both of them are made from high quality grain that endowed by nature, which makes them *take the lead* in the Baijiu field. Through the two *representative* brands, not only do we enjoy Baijiu, but we are also able to appreciate thousands of years of national traditions. Culture of the two brands originate from the idea of "valuing harmony and the *doctrine* of the mean" in the *Book of Changes*.

The culture of Wuliangye is developed on the basis of the doctrine of the mean in the Confucian culture, that is, pursuing the harmonious unity between the tradition and modernity. From the past to the present, Wuliangye has been advancing the development of its culture under the unique natural *ecological* environment, with the unique ancient cellar of the Ming Dynasty (638), the exclusive *formula* of five grain and the special brewing technology. Wuliangye pays close attention to the senses of color, *fragrance* and taste, especially the *inherent* quality of Baijiu. Meanwhile, Wuliangye gives priority to the aftertaste. There are both strength and grace in its quality, which endows Wuliangye with rich taste and full flavor, and brings the three aesthetic senses into full play and achieves the mean and harmony. Besides inheriting the Confucian doctrine of the mean, Wuliangye has also evolved into a vigorous and enterprising culture in the long history. It could be felt from the law-of-the-jungle theme sculpture in the Wuliangye Century Square. The moment people enter the Group of Wuliangye, they can make sense of the cruel market competition as well as the *doctrine* "The strong survive, the weak perish". As a result, Wuliangye has to strive and thrive to earn more developmental room.

Different from Wuliangye, the culture of Moutai has gradually developed into the Taoist culture—doing nothing if it goes against the nature; in other words, everything should follow the rules of nature. The trademark of Moutai is associated with the sea, which could *accommodate* and undertake the spirit of the universe. Moutai thinks highly of a kind of static and peaceful culture which could be seen from its totems. Moutai has taken the idea of inaction in the Taoist culture as its unique spirit. Its brewing methods of *open fermentation* and *closed fermentation* corresponding to the concept of Yin and Yang in Tai Chi. In the market competition, Moutai has held the banner of "brewing national Baijiu" for a long time in order to call on the Baijiu market. Moutai has been committed to Baijiu quality for many years and advocated the healthy quality to form an irresistible trend. In order to take the lead in the fierce competition, Moutai insists on the good quality as well as its own culture. It mainly includes four commandments: loyalty, *filial* piety, integrity and faithfulness. Loyalty refers to the idea that people should work hard for their country and bring glory for it; filial piety holds that people should obey their parents and sacrifice for their family; integrity argues that people should not change his position for money; and faithfulness stresses that people should not persuade others to drink too much. The four commandments are called national Baijiu culture of Moutai.

Although there are some differences between the two famous Baijiu brands, both of them are beloved by consumers. What's more, both of them are inheriting and *reviving* traditional Chinese culture. Nowadays, Wuliangye and Moutai are striving to go abroad in order to make "Chinese-famous Baijiu" become "world-famous Baijiu". In the process, both of them will make more efforts to introduce Chinese culture to the world so that it can be better understood and accepted.

New Words and Expressions

1. take the lead 带头的
2. representative [ˌreprɪˈzentətɪv] *adj.* 具有代表性的
3. doctrine [ˈdɒktrɪn] *n.* 学说
4. ecological [iːkəˈlɒdʒɪkəl] *adj.* 生态的
5. formula [ˈfɔːmjələ] *n.* 配方

6 fragrance [ˈfreɪgrəns] *n.* 芳香

7 inherent [ɪnˈherənt] *adj.* 内在的

8 accommodate [əˈkɔməˈdeɪt] *v.* 容纳

9 open fermentation 开放式发酵

10 closed fermentation 封闭式发酵

11 unparalleled [ˌʌnˈpærəleld] *adj.* 无与伦比的

12 filial [ˈfɪlɪəl] *adj.* 孝顺

13 revive [rɪˈvaɪv] *v.* 复苏

Exercises

I. Choose the best answer to each of the following questions or incomplete statements.

1 Which group is the most famous brand in China?
   A. Wuliangye and Tuopai.
   B. Moutai and Wuliangye.
   C. Langjiu and Moutai.
   D. Wuliangye and Swellfun.

2 The two Baijiu cultures originate from _____.
   A. *the Book of Changes*
   B. *the Book of Mencius*
   C. *the Book of Xunzi*
   D. *the Book of Confucius*

3 Both of the two Baijiu cultures represent the thought about _____.
   A. harmony   B. changes
   C. development   D. nothing

4 Wuliangye was invented in _____.
   A. the Qing Dynasty   B. the Ming Dynasty
   C. the Tang Dynasty   D. the Song Dynasty

UNIT 2 Legends and Culture of Liquor in China

5 Wuliangye pays attention to _____.
   A. the senses of color, fragrance and taste
   B. the senses of color, touch and tastes
   C. the senses of color, feeling and tastes
   D. the senses of color, sight and tastes

6 Moutai appreciates the _____ culture.
   A. quiet and peaceful
   B. static and quiet
   B. static and peaceful
   D. noisy and peaceful

7 The national culture of Moutai includes _____.
   A. loyalty, filial piety, temperateness and justice
   B. mean culture, filial piety, temperateness and justice
   C. loyalty, filial piety, quietness and justice
   D. loyalty, filial piety, temperateness and injustice

8 What is the plan of the two famous Baijiu brands in the near future?
   A. Sell more Baijiu.
   B. Introduce new Baijiu.
   C. Purchase foreign wine.
   D. Introduce Chinese culture with Baijiu.

II. Translate the following passage into English.

中国人在七千年以前就开始用谷物酿酒。总的来说，不管是古代还是现代，酒都和中国文化息息相关。长久以来，酒文化在中国人生活中一直扮演着重要的角色。我们的祖先在写诗时以酒助兴，在宴会中和亲朋好友敬酒。作为一种文化形式，酒文化也是普通百姓生活中不可分割的一部分，比如在生日宴会、送别晚宴、婚礼庆典等场合，人们都会举杯邀酒，表达情谊。

UNIT 3
Drinking and Health

Preview

"*Jiu*", which refers to all kinds of alcoholic beverages in Chinese, is a kind of ancient drink with far-reaching youthful flavor, and is one of the products and symbols of human material civilization. All alcohol beverages, including Baijiu, beer, wine and etc. are collectively known as liquor. China, as one of the first countries in the world to make liquor, is also the empire for drinking. Drinking has both advantages and disadvantages, which mainly related to the amount of drinking and the type of liquor. Moderate drinking is beneficial to health, while long-term heavy drinking is harmful to human body.

Section A

Benefits of Drinking

The word "alcohol" is *derived* from the Arabic word "al kuhul", which originally referred to a powder used by women for make-up, and then it referred to "the essence of wine", known as alcohol. In general, liquor, in addition to *ethanol*, contains *esters*, acids, *phenols* and amino acids and other substances; What's more, it is made from grain or fruits. Therefore, liquor, which is spicy, sweet, and hot, has the essence of grain and water. It is easy to enter the two *meridians* of heart and liver, thus, it has the effect of *unblocking* blood vessels, promoting Qi (which refers to the power that enables human organs to function properly in the traditional Chinese medicine) and blood circulation, *dispersing* wind- and accumulated-cold, curing stomach cold, strengthening *spleen* and stomach, and enhancing efficacy of medicine. Drinking moderately can not only make people think actively and stimulate people's wisdom, but still can strengthen the mind, eliminate fatigue and promote sleep. When alcohol enters the body, it expands blood vessels and increases blood flow. Baijiu is a *stimulus* to taste and smell, which can inturn increase breathing volume and promote appetite. Tests have shown that a small amount of alcohol in human body can increase the content of HDL (high density *lipoprotein*) in the blood and reduce the level of LDL (low density lipoprotein). Thus, a small amount of alcohol can reduce the chance of *atherosclerosis* and obstruction caused by fat deposition. Moderate liquor consumption can promote blood circulation and reduce the possibility of death from heart disease. Alcohol actually helps prevent *thrombosis*, thereby effectively preventing stroke and heart attacks. People who drink moderately even live longer than those who never drink!

Drinking can be beneficial in the following aspects.

1 Liquor as a carrier of emotion

Drinking could make people calm and cheerful, which is a spiritual force of drinking that has been passed down from ancient times. Throughout China's ancient and modern drinks, the cultural effects of liquor are remarkable: when one feels happy, he always drinks "grape wine with *luminous* jade cup"; when *decadent*, "enjoy while one can"; when missing relatives and friends, "How long will the full moon appear? Wine cup in hand, I ask the sky"; when gathering with friends, "when one drinks with a good friend, a thousand cups are not enough"; when feeling lonely, "I raise my cup to invite the Moon who blends Her light with my Shadow and we're three friends"; when saying goodbye, "I invite you to drink a cup of wine again; West of the Sunny Pass no more friends will be seen". It can be said that liquor can both make merry and relieve worry. Liquor *permeates* every corner of Chinese social life and becomes a carrier of culture, which is known as "spirits culture" and also adds many colors to human cultural life. As the saying goes, "No liquor, no banquet." Liquor adds a lot of topics to the dinner, for instance, while people are drinking and talking in the atmosphere of liquor, they are cheerfully chatting and laughing, which can warm the whole body. If you are in a good mood, you will naturally be healthier. Of course, if you drink too much, it will not only hardly relieve your worries, but more often affect your health.

2 Liquor as medicine

Liquor can not only carry emotions, but also help cure disease and nourish us. Liquor is a "life-saving medicine", but sometimes it is also a "killing weapon". Poisonous liquor, such as Zhenjiu, can put people to death. Liquor can be used in medicine because it is a good *solvent* which can dissolve many poorly soluble or even indissoluble substances. Besides, it is used to make *pharmaceutical* liquor, to *decoct* traditional Chinese herbs with or better effect than water. In addition, as the pharmaceutical liquor is absorbed after getting into the body, it immediately flows into the blood, which has a greater efficacy, thus playing a *therapeutic* tonic effect. For this reason, Chinese medicine prescriptions often suggest patients to drink with liquor or use liquor as a guiding drug when decocting herbs. Liquor is not only taken orally, but also for surgery. In addition to alcohol *disinfection*, liquor, such as tiger-bone liquor and Shiguogong liquor, can be applied to the affected area for treatment of sprain, arthritis, nerve numbness. Hua Tuo, a famous doctor of the Eastern Han Dynasty, once used liquor to dissolve a drug called "Mafeisan" (powder for *anesthesia*) to complete surgery as an *anesthetic* successfully. This was the first of its kind in the world in the 2nd century A.D. Although it is no longer in use, liquor played a role in helping patients undergoing surgery in an age when anesthetics were not invented. Different alcoholic beverages have different medical effects. According to the latest research, drinking one or two cups of strong alcoholic beverages per day can reduce the incidence of coronary heart disease, thereby reducing the risk of death from coronary heart disease. Red grape wine has been selling well in China in recent years, because moderate drinking can not only prevent aging, but also can prevent diseases caused by aging. Red grape wine has a *polyphenolic* chemical called anthocyanidin, which is the source of red and has a strong *antioxidant* function. If you can resist oxidation, you can resist aging. Anthocyanins are found mainly in the skins and seeds of red grapes and are converted into antioxidants during the brewing process.

3 Liquor and fitness

The effect between liquor and bodybuilding can be traced back to Su Jing's book entitled *New Revised Herbs* as early as in the Tang Dynasty, in which it writes, "liquor, which refers to red grape wine, can warm kidney, retain youthful looks, and resist cold". In the middle of the seventh century, grape wine was introduced into China and saw development. Peach-blossom liquor is also worth mentioning. It is made by dried peach blossom picked in March, then soaked in the first-class Baijiu, and stored for 15 days. It has the effect of skin-moisturizing and blood-activating, preferred by people who wish to stay young and beautiful. Other liquor such as white pigeon liquor, Longan liquor and Sparkling liquor also have *cosmetic* effects. In order to make hair and skin strong and handsome, ancient Chinese people had the practice of bathing with liquor. Liquor bath also prevails in Japan. Before entering the bath, one will prepare 0.75kg of liquor named "jade skin" for both drinking and bathing. After the bath, the skin will be as fair as jade and the body warm all over. Jade skin for bath is a kind of saki made from the steamed mixture of fermented grain. According to a survey conducted by medical experts, liquor has a benign stimulation to the skin, which can accelerate blood circulation and is of great benefit to the body. Saki bath is popular in Japan.

4 Liquor and cooking

People often use moderate amount of liquor when cooking, which can remove fishy smell, make dishes sweet and delicious. Because the main ingredient of liquor is ethanol, whose boiling point is low, it becomes very volatile once heated, which takes away the smell of fish, meat and the likes. Millet wine, between beer and Baijiu, is the most ideal cooking liquor, because it contains moderate ethanol content, and it is rich in amino acids, which can produce *amino acid sodium salt* in the cooking process with salt, namely *monosodium glutamate*, which can increase the delicacy of dishes. In addition, as the millet wine is equipped with aromatic Chinese herbal medicines, the dishes will have a special aroma when it is used as sauce. Of course, in the absence of millet wine, other liquor can also be used instead.

New words and Expressions

1. derive [dɪˈraɪv] v. 源于，来自
2. ethanol [ˈeθənɒl] n. 乙醇
3. ester [ˈestə(r)] n. 酯
4. phenol [ˈfiːnɒl] n. 苯酚，石碳酸
5. meridian [məˈrɪdiən] n. 针灸经络
6. unblock [ˌʌnˈblɒk] v. 疏通
7. disperse [dɪˈspɜːs] v. 分散；散开；驱散
8. spleen [spliːn] n. 脾
9. stimulus [ˈstɪmjʊləs] n. 刺激；刺激品
10. lipoprotein [ˌlɪpəʊˈprəʊtiːn] n. 脂蛋白
11. atherosclerosis [ˌæθərəʊskləˈrəʊsɪs] n. 动脉硬化，动脉粥样硬化
12. thrombosis [θrɒmˈbəʊsɪs] n. [医] 血栓形成；血栓症
13. luminous [ˈljuːmɪnəs] adj. 明亮的；发光的
14. decadent [ˈdekədənt] adj. 颓废的；衰落的
15. permeate [ˈpɜːmɪeɪt] v. 弥漫；渗透
16. solvent [ˈsɒlvənt] n. 溶剂
17. pharmaceutical [ˌfɑːməˈsjuːtɪkəl] adj. 制药的；配药的
18. decoct [dɪˈkɒkt] vt. 熬；煎
19. therapeutic [ˌθerəˈpjuːtɪk] adj. 治疗的
20. disinfection [ˌdɪsɪnˈfekʃən] n. 消毒；灭菌

21 anesthesia [ˌænɪsˈθiːziə] *n.* 麻醉
22 anesthetic [ˌænɪsˈθetɪk] *n.* 麻醉剂，麻醉药
23 polyphenolic [ˌpɒlɪfɪˈnɒlɪk] *adj.* 多酚的
24 anthocyanidin [ˌænθəsaɪˈænɪdɪn] *n.* 花色素
25 antioxidant [ˌæntɪˈɒksɪdənt] *n.* 抗氧化剂，硬化防止剂
26 cosmetic [kɒzˈmetɪk] *adj.* 化妆用的；美容的
27 amino acid *n.* 氨基酸
28 sodium salt *n.* 钠盐
29 monosodium glutamate 味精

Exercises

I. Answer the following questions.

1 Where is the word "alcohol" derived from?
2 How do people express their feeling with liquor from ancient times?
3 How do you understand the proverb "No liquor, no banquet."?
4 Why can liquor be used in medicine?
5 What effect does the anthocyanin contained in red wine have?
6 What is the effect of peach-blossom liquor?
7 What is the benefit of liquor bath?
8 Why is millet wine the most ideal cooking liquor?

II. Discuss the following questions.

1 Give a brief introduction of the relationship between liquor and health.
2 Tell your partner some interesting stories about drinking.

III. Match the items in column A with those in column B.

A	B
1. 酒之精华	____a. polyphenolic chemical
2. 祛风散寒	____b. When one drinks with a good friend, a thousand cups are not enough.
3. 脂肪沉积	____c. No liquor, no banquet.
4. 血液循环	____d. make pharmaceutical liquor
5. 夜光杯	____e. a life-saving medicine.
6. 酒逢知己千杯少	____f. dispersing wind-cold
7. 无酒不成席	____g. fat deposition
8. 救人的良药	____h. luminous jade cup
9. 泡制药酒	____i. blood circulation
10. 多酚的化学物质	____j. the essence of liquor

IV. Fill in the following blanks with the words given below. Change the form when necessary.

| derive | unblock | nocturnal | stimulus | decadent |
| permeate | decoct | cosmetic | therapeutic | disperse |

1 If your wine is still, pour it into the glass to let the bouquet _____ the vessel.
2 The two attitudes _____ from different historical perspectives.
3 Have you ever _____ Chinese medicinal herbs?
4 The wind _____ the cloud from the sky.
5 Her words of praise were a(n) _____ for people to work harder.
6 Mourning can have a(n) _____ function that we ignore at our peril.
7 I don't like the _____ opinion in the article.
8 The _____ brand is her favorite.
9 We had to call a plumber to _____ the drains.
10 _____ animals such as bats and owls only come out at night.

V. Translate the following passage into English.

酒和饮酒文化在中国的历史中占据着重要地位。从宋代开始，白酒成为中国人饮用的主要酒类。中国白酒制作工艺复杂，原料丰富多样，是世界著名的六大蒸馏酒（distilled Baijiu）之一。中国有很多优秀的白酒品牌，受到不同人群的喜爱。在当代社会，饮酒文化得到了前所未有的丰富和发展。不同地区和场合的饮酒习俗和礼仪已成为中国人日常生活中重要的组成部分。在几千年的文明史中，酒几乎渗透到社会生活中的各个领域，如文学创作、饮食保健等。

VI. Write a passage according to the given outline with no less than 150 words.

Excessive Drinking（过度饮酒）

1) 过度饮酒的现象；
2) 过度饮酒带来的危害；
3) 为了保持身体健康，我们应该怎么做。

Section B

How to Drink in a Healthy Way

[A] Baijiu is quite common in people's life, but excessive drinking is harmful to health. If you want to keep fit through drinking, you must know how to drink healthily, and understand what should not be eaten while you are drinking.

[B] In today's society, we often hear the news of people who are poisoned, even killed by alcohol, so we advise people to drink in a healthy way. Here're some tips that benefit you most.

1 Add some soda to the liquor.

[C] Soda that is added in the liquor can reduce the amount of alcohol that your body absorb, so you won't get drunk and red-faced right away.

2 Do not drink on an empty stomach.

[E] It is better to eat while drinking, to make liquor stay in the stomach as long as possible. This is an easy way to take to avoid getting drunk.

[F] One of the characteristics of alcohol is that its absorption in the digestive organs is very fast. Of alcohol entering the body, 20% is absorbed by the stomach, while 80% by the small *intestine*, which can be dissolved in the blood and be transported to every corner of the body.

[G] If there is food in the stomach, the movement of alcohol to the small intestine will slow down; if there is nothing covering the gastric mucosa on an empty stomach, alcohol will pass *unimpeded* through the stomach and go straight to the small intestine, which will *accelerate* the absorption rate of alcohol, make the alcohol concentration in the blood rise sharply, putting yourself into a dangerous state of *paralysis* dramatically in an instant.

3 Do not drink too fast.

[H] Drink slowly, and pause from time to time to take some food. Do not drink carbonated drinks, such as Sprite and Coke, etc. while having a drink, so as not to speed up the body's absorption of alcohol.

4 Do not choose the wrong sobering.

[I] Tea is traditionally regarded as a good sobering, and even Sprite or Coke is chosen to sober up. However, the best choice to ease hangover is fruit juice, especially orange juice, which can play a good role in antialcoholic, because juice contains *fructose*, which can help alcohol burn effectively.

5 Do not drink different liquor at the same time.

[J] Different alcoholic beverages vary in composition and content. They will undergo reactions if mixed with each other, which will make people easy to get drunk or feel uncomfortable after drinking, and will even cause headache.

6 Do not take a shower after drinking.

[K] If people bathe immediately after drinking, the glucose stored in the body will be consumed quickly, thus resulting in a decrease in blood sugar content and a sharp drop in body temperature. Alcohol can inhibit the normal activity of the liver and hinder the recovery of glucose storage in the body, thus endangering life and possibly causing death.

[L] Besides the tips mentioned above, there are also some wrong ideas about drinking.

7 Eat meat but no staple food while drinking.

[M] This equals to *chronic* suicide, as there is little nutrition in liquor, but calories. If you always drink without having staple food, it can not only cause indigestion, but also *malnutrition*. Eating meat but no staple food while drinking will do harm to you. First, you are likely to fall into malnutrition, since 70% of human nutrition comes from staple food, while the rest from non-staple food. Second, regular drinkers often suffer a lot from the deficiency of vitamin B. *Enzymes* in the body become inactive without enough intake of vitamin B contained in staple food, which would cause various malfunctions of the body.

8 Drinking does not promote sleeping.

[O] It is traditionally believed that drinking can make people drowsy, but it is not the case. A small amount of alcohol can make people excited, whereas a large amount of drinking will cause depression, leading to shallower breath, lower *metabolism*, and *hypothermia*.

[P] Many parts of the body will be *congested* if drink too much. For example, throat congestion can make one snore while sleeping, and *flabby* muscles can even make breathing hindered or cause respiratory obstruction, which eventually lead to *cardiovascular* and *cerebrovascular* diseases or accidental death.

New Words and Expressions

1. gastric [ˈgæstrɪk] *adj.* 胃的；胃部的
2. intestine [ɪnˈtestɪn] *n.* [解] 肠
3. unimpeded [ˌʌnɪmˈpiːdɪd] *adj.* 畅通无阻的
4. accelerate [əkˈseləˌreɪt] *vt.* （使）加快，（使）增速
5. paralysis [pəˈræləsɪs] *n.* 麻痹，瘫痪
6. sober [ˈsəʊbə(r)] *v.* （使）冷静，（使）清醒
7. antialcoholic [ˌæntɪˌælkəˈhɒlɪk] *n.* [医] 戒酒药
8. fructose [ˈfrʌktəʊs] *n.* 果糖
9. glucose [ˈgluːkəʊs] *n.* [化] 葡萄糖，右旋糖
10. inhibit [ɪnˈhɪbɪt] *v.* 抑制；禁止
11. chronic [ˈkrɒnɪk] *adj.* 慢性的；长期的
12. malnutrition [ˌmælnjuːˈtrɪʃn] *n.* 营养不良
13. enzyme [ˈenzaɪm] *n.* [生化] 酶
14. metabolism [mɪˈtæbəlɪzəm] *n.* 新陈代谢；代谢作用
15. hypothermia [ˌhaɪpəʊˈθɜːmɪə] *n.* 低体温
16. congest [kənˈdʒest] *vt.* 充血
17. flabby [ˈflæbɪ] *adj.* 肌肉松垂的
18. cardiovascular [ˌkɑːdɪəʊˈvæskjʊlə] *adj.* 心血管的
19. cerebrovascular [ˌserɪbrəʊˈvæskjʊlə] *adj.* 脑血管的

Exercises

I. Reading comprehension.

The following ten statements are derived from the passage. Each statement contains information given in one of the paragraphs. Identify the paragraph from which the information is derived. Each paragraph is marked with a letter.

1. A small amount of alcohol can make people excited, but drinking too much will cause depression.

2 When drinking, you'd better not drink carbonated drinks, such as Cola, Sprite, etc.
3 Because the various alcoholic beverages contain different composition and contents, you'd better not mix them with each other, or you may feel uncomfortable after drinking.
4 As the drink made with *Gegen* can relieve a hangover, so people can drink it to sober up.
5 If you drink several cups of liquor in succession or with no food, the alcohol density in the blood will rise dramatically.
6 Tea is generally believed as the good sobering, however, the best choice to relieve the hangover is fruit juice.
7 Adding soda to liquor is advised.
8 After drinking, the glucose stored in the body will be depleted by physical activity in the shower, leading to a decrease in blood sugar content and a sharp decrease in body temperature.
9 One of the features of alcohol is that its absorption in the digestive organs is very quick
10 Always eating no staple food like rice while drinking, one is inclined to fall into malnutrition.

II. Translate the following passage into English.

在中国，酒作为一种特殊的文化形式，有着五千多年的历史。林超所著的《杯里春秋》(*The Spring and Autumn in the Cup*)一书认为，喝酒有点像做学问，而不是大吃大喝。中国历史上有很多关于酒的故事。唐代伟大诗人李白可以"斗酒诗百篇"，喝得越多，他的诗就作得越好。在中国民俗中，酒有着极其重要的地位。不论是君王还是平民，都会饮酒来庆祝节日、婚礼、生日聚会，纪念逝者，为亲友接风或送行，庆贺好消息，摆脱焦虑，甚至治疗疾病以求长寿。

UNIT 4

Baijiu Map

Preview

As an essential component of Chinese diet culture, Baijiu occupies a highly important position in the life of Chinese people. The tradition of brewing Baijiu maintained from ancient times to the present all around China. But miraculously, the regionality determinates the local Baijiu quality. Plant constitution in different regions decides the different choice of raw materials for making Baijiu, and the local ecological conditions affect the intrinsic temperament of Baijiu in the region. Furthermore, the preference for flavor and taste of Baijiu varies because of the regionality and ethnicity. Distinctive region characteristics construct a basis for unique Chinese Baijiu culture.

Section A

Main Distribution of Baijiu and the Most Representative Brands of the Regions

The unique features of a local environment give special characteristics to its inhabitants. Also, the features of a local environment determine the peculiarity of spirits which can *differentiate* from that made in other areas. Undoubtedly, each brand can be regarded as the particular speciality of the place of origin, it is most especially so for traditional Chinese Baijiu.

In China, Baijiu (also called Shaojiu or Baigan) is a time-honored spirit with a history of more than 3,000 years. Depending highly on environmental factors such as water, soil, climate, temperature and microbe, Baijiu *is endowed with* distinct attributes by localities in different geographical conditions and folk customs, which is known as "region characteristic". There exists three major producing areas in China according to the different geographic factors and production techniques which can influence the process of Baijiu-producing and the current productivity and output of the Baijiu enterprises: Chuan-Qian (Sichuan province and Guizhou province) production area, Huang-Huai (the Yellow River and the Huai River Valleys) production area and Lianghu (Hunan province and Hubei province) production area.

Chuan-Qian production area

What is universally recognized is that as the *ancestral* lands of the ancient and mystic Bo nationality, the region of Sichuan province and Guizhou province is the *cradle* of the best Baijiu of the Luzhou-flavor-Baijiu and Moutai-flavor-Baijiu, it is also the *epitome* of Chinese Baijiu culture.

As early as the pre-Qin period, Bo people's technology of making *betel pepper* (a kind of fruit wine) had been matured. In the late 3,000 years, the formula of Baijiu and the process of making Baijiu have been innovating and perfecting *unremittingly*. *Endeavoring* to preserve the original ecological culture of Baijiu and maintain sustainable and healthy development of the Baijiu brewing industry, most *distilleries* in Yangtze River Valley, Minjiang River Valley and the Chishui River Basin still stick to the traditional Baijiu brewing technology nowadays. *Dotted with* alcohol-producing cities, distilleries and taverns, this region accumulated swarms of Baijiu brands and series of Baijiu products, best representing the rich heritage of Chinese Baijiu culture. It has become the core of creating the world-class regional Baijiu brands in China.

Chuan-Qian production area gathered a large collection of brands of traditional distillated Baijiu. China Golden Triangle of Baijiu, consisting of Yibin, Luzhou and Zunyi, has bred many famous international Baijiu brand such as Wuliangye, Moutai, Jiannanchun, Luzhou Laojiao, Tuopai, Swellfun and Langjiu. Baijiu yield in this region has accounted for more than 20% of the gross Baijiu output in China. Known as "the biggest *industrial cluster*, the centralized brand cluster, the most efficient capacity cluster and the greatest policy preference", the region was honored as "the best *ecotope* which is suitable for brewing high-quality pure *distilled* Baijiu on the same *latitude* of the earth" by *FAO*.

Huang-Huai production area

As one of the birthplaces of Chinese civilization, the Yellow River and Huai River Valleys fostered lots of outstanding talents and had witnessed prosperity in society, economy and culture as never before since the Qin and Han Dynasties. Especially from the period of Three Kindoms Dynasties (220-265 A.D.) to the Tang and Song Dynasties (960-1279 A.D.), Huang-Huai area had turned into a critical region of the inheritance of Chinese civilization.

Huang-Huai production area locates between Yangtze River and Yellow River Valleys, it is not only influenced by the two river systems, but also affected by geographical conditions and the *monsoon* climate of medium latitudes. The distinctive climate, huge grain production and the *irreproducible* geographic location of north-south crossroads make it a treasure to brew elegant-type Luzhou-flavor-Baijiu. Baijiu that conform to the regional features and humanistic characters of Huang-Huai area, including Yanghejiu of Jiangsu province, Gujingjiu of Anhui province, Dukangjiu and Songhejiu of Henan province and Confucius Family Liquor of Shandong province, not only *complement each other with* the history and humanistic characters of Huang-Huai area, but also make the Huang-Huai Baijiu zone some kind of phenomenon all along.

The strategic alliance of Baijiu enterprises from Henan, Shandong, Jiangsu and Anhui provinces was made in 2007. Especially the transition of Baijiu style from focusing on *bouquet* to taste made Huang-Huai production area a significant force in Chinese Baijiu *domain*. In 2015, Huang-Huai Famous Baijiu Development League was born, which aims to run the brands bigger and stronger, forge the unified taste and style, improve brands' impact and exploit markets as a whole. The establishment of the League will drive the products' sustained innovation, increase brands awareness and boost sales effectively.

The Lianghu (Hunan and Hubei provinces) production area

Lying in the middle reaches of the Yangtze River, the Lianghu area is a main base of agricultural production and developed economic zone in China's history and *is renowned as* "the lands of fish and rice". This region ranks very high in Baijiu production areas because of abundant water resources and Baijiu-making grain.

Intersecting in the Lianghu production area, Luzhou-flavor Baijiu belt from Sichuan to Anhui and Jiangsu provinces and Moutai-Luzhou-flavor Baijiu belt from Shanxi province to Guangdong and Guangxi regions meet in the *prefecture-level cities* Jingzhou, Yichang and Changde. Therefore, Baijiu flavor types including Moutai-flavor, Fen-flavor, Lu-zhou-flavor, special-flavor, heavy-perfume-flavor and so on are plentiful because of the unique natural environment and microbiological conditions. The top-ranked nationwide Baijiu yield, high degree of industrial cluster together with excellent sales achievements qualified this region as the third largest one after Chuan-Qian production area and Huang-Huai production area in China. For the past few years, Baijiu brands registered by enterprise in this region such as Daohuaxiangjiu and Zhijiangjiu of Hubei province and Liuyanghejiu and Jiuguijiu of Hunan province have gained spectacular popularity and sales performance.

New Words and Expressions

1 differentiate [ˌdɪfəˈrenʃɪeɪt] *vt. & vi.* 区分，区别
2 be endowed with 被赋予
3 ancestral [ænˈsestrəl] *adj.* 祖先的；祖传的

4 cradle [ˈkreɪdl] *n.* 摇篮；发源地
5 epitome [ɪˈpɪtəmi] *n.* 缩影
6 betel pepper 蒟酱，一种果酒
7 formula [ˈfɔːmjələ] *n.* 配方
8 unremittingly [ʌnrɪˈmɪtɪŋli] *adv.* 不懈地；不间断地
9 endeavor [ɪnˈdevə] *vi.* 努力；尽力
10 distillery [dɪˈstɪləri] *n.* 酿酒厂；蒸馏室
11 dot with 散布
12 industrial cluster 产业集群
13 ecotope [ekəˈtəup] *n.* [生态] 生态区
14 distill [dɪsˈtɪl] *vt.* 提取；蒸馏
15 latitude [ˈlætɪtjuːd] *n.* 纬度；纬度地区
16 FAO abbr.（Food and Agriculture Organization）联合国粮食与农业组织
17 monsoon [mɒnˈsuːn] *n.* 季风
18 irreproducible [ˌɪrɪprəˈdjuːsɪbl] *adj.* 不能复制的
19 bouquet [bəʊˈkeɪ] *n.* 酒香
20 domain [dəˈmeɪn] *n.* 领域
21 prefecture-level cities 地市级城市
22 complement each other with 相得益彰
23 be renowned as 因……而著名

Exercises

I. Answer the following questions.

1 What are the three major production areas of Baijiu?
2 What are the factors that influence the production of Baijiu?
3 Who developed the technology of Baijiu-making?
4 What makes the China Golden Triangle of Baijiu the core of Baijiu-making?
5 What effect will the strategic alliance of Baijiu enterprises from the four provinces have?
6 What advantages do the provinces Hunan and Hubei have in Baijiu-making?

7 Why does the Lianghu production area have such plentiful flavor types?

8 Why does drinking Baijiu become a form of culture in China?

II. Discuss the following questions.

1 Talk about the inheritance of technical elements of solid liquor.

2 Discuss other Baijiu production areas.

III. Match the items in column A with those in column B.

A	B
1. latitude	_____ a. 配方
2. flavor type	_____ b. 产业集群
3. Moutai-Luzhou flavor	_____ c. 纬度
4. betel pepper	_____ d. 白酒金三角
5. heavy-perfume-flavor	_____ e. 浓香型
6. distill	_____ f. 酒香
7. China Golden Triangle of Baijiu	_____ g. 浓酱兼香型
8. formula	_____ h. 香型
9. bouquet	_____ i. 提取，蒸馏
10. industrial cluster	_____ j. 蒟酱

IV. Fill in the following blanks with the words given below. Change the form of the words when necessary.

innovation	distill	foster	forge	endow	differentiate
sustain	domain	endeavor	complement		

1 Its unusual nesting habits _____ this bird from others.

2 In the long term, growth in emerging economies will be contingent on their ability to _____ and shift away from an externally driven growth model to one fueled in part by local consumption.

3 The government should do more to promote _____ agriculture.

4 The quality features of accounting information should be _____ with new intension and added with new meaning.

5 We cannot succeed in this _____ alone, but we can lead it.

6 Back in the 1980s, they were attempting to _____ a new kind of rock music.

7 If you do not have _____ knowledge of the business area, try to work with people that can provide this knowledge.

8 Only when the diversity of the world is respected can various ethnic groups and civilizations live in harmony, learn from each other and _____ each other.

9 Having limited space to make your point really forces you to _____ and clarify your message.

10 He said that developed countries had a responsibility to _____ global economic growth to help new democracies.

V. Translate the following passage into English.

On September 19, 2010, with a total investment of about 5 billion yuan, the project of Wuliangye Cultural Blocks in Yinbin the city of liquor in China Golden Triangle of Baijiu that covers an area of over 250 hectares started. Wuliangye Cultural Blocks is the most invested project ever in Sichuan province with spirits culture as its theme. It is divided into two areas, North and South. The ecological character of the ten-li Baijiu city and the Jiuzhou Tower park will fully display the Sichuan liquor; ancient cellars such as Lichuanyong will be designed and protected as historical sites of the city of liquor, which will be regarded as the source of liquor in Yinbin. Through the planning and construction of projects like Baijiu culture museum and the Liubeichi Baijiu culture theme park, the 4,000-years brewing history of Sichuan Baijiu represented by Yinbin the city ot liquor will be presenting to the public.

VI. Write a passage according to the given outline in no less than 150 words.

The Creation of the China Golden Triangle of Baijiu

1) 四川盆地具有独特的酿酒优势环境；
2) "白酒金三角"的建立有助于将优势最大化；
3) 企业发展与生态保护共生共荣。

Section B

China's *Bordeaux*: the Golden Triangle of Baijiu on the Upper Reach of the Yangtze River

In the border region of Southwest China, three major cities of Yibin, Luzhou and Zunyi in the junction of Guizhou and Sichuan Provinces—home to some of the country's *landmark* liquor brands, such as Moutai, Wuliangye, Langjiu and Luzhou Laojiao—are now building up in the region to create a Golden Triangle of Liquor, as the Chinese equivalent of the French Bordeaux region.

■ Good water brewing good liquor

Although grain liquor, or Baijiu, is produced throughout China, the Triangle region accounts for the largest proportion of Baijiu production facilities and distilleries.

Situated in the Sichuan Basin, the region's *subtropical* climate, humid weather, clean water, rich soil and basin landscape create ideal environment for brewing good Baijiu made from *sorghum*, wheat or rice, forming centuries-old Baijiu culture and brewing tradition.

As the Chinese saying goes, good liquor is brewed where there is good water. Since the Tuo River, Chishui River, Min River and the Yangtze River all flow through the Triangle area, its abundant and clean water resource serves as fundamental material for liquor making in the area.

"This natural environment is a special ecosystem with various contributing factors and therefore is *unduplicated* and inimitable," said Peng Zhifu, deputy chief manager of liquor giant Wuliangye Group Co. Ltd. in Yibin.

"Moutai would not have existed without the healthy ecosystem along the Chishui River. The river is Moutai's life," agreed Li Baofang, chairman and general manager of Moutai Group in Guizhou province. Bacterial and *archaeal* diversities in Moutai section of the Chishui River and the grain cultivated with the river water produce the unique flavor of Moutai.

The precious ecosystem that breeds good liquor is also under careful protection by these companies along with local authorities.

In an effort to protect the ecosystem along the river, Moutai company has invested 468 million yuan to build five *sewage-processing plants* with an annual capacity of over two million tons.

Since 2006, Luzhou began to build the China Golden Triangle Baijiu Industry Park, inviting in qualified upstream and downstream enterprises related to the Baijiu industry. The industrial cluster enjoys joint and standardized improvements in sorghum planting, Baijiu brewing, Baijiu storage, packaging, supply and *logistics distribution*. Meanwhile, by providing *infrastructure* and supporting services such as sewage-processing plants and national-standard *quality supervision* center at the industry park, it also helps guarantee that the whole chain is eco-friendly.

For the local cultivation of high-quality *glutinous* Southern Red Sorghum, a particular grain used as the raw material to make Baijiu, currently major liquor companies such as Luzhou Laojiao ("laojiao" means old cellar), a well-known Luzhou-based Baijiu producer, have started to build their own *organic* raw material base, a move not only to help protect the quality of Baijiu products, but also to promote local agricultural industry, helping farmers to increase their incomes.

Spirits culture, another business card

The history of Baijiu brewing along the Yangtze River dates back to centuries ago when history records the Tang Dynasty (618-907 A.D.), while the nation's drinking history can be traced back even further to the Shang Dynasty (1600-1046 B.C.).

Baijiu, the most famous kind of liquor in China, rules almost every festive occasion in China, where it's the *tipple* of choice for everything from wedding receptions to business banquets. Many ancient Chinese poets such as Li Bai and Du Fu also had their legendary life and all-time popular works associated with drinking traditional Baijiu.

The unique spirits culture has made the companies' brewing techniques and materials a must-see for tourists traveling in the region. They are also listed as national *intangible* cultural heritage.

Luzhou Laojiao Group is one of the oldest producers in the Triangle region, whose biggest attraction is its legendary liquor cellars, which dates back to the year of 1537.

Using techniques and even brewing pits that originated in the Ming (1368-1644 A.D.) and Qing (1644-1911 A.D.) Dynasties, the brand continues to make Baijiu the same way 23 generations later.

"The older the cellar is, the better the brewed liquor is," said Zeng Na, the 23rd generation inheritor of Baijiu tasting at Luzhou Laojiao.

By 2013, the city of Luzhou has 1,615 century-old brewing cellars, 16 Baijiu workshops which date back to the Ming and Qing Dynasties, and 3 natural liquor caves, making the city a special tourism destination for spirits culture experience.

Baijiu going global

Although it is the most widely drunk hard liquor in the world, Baijiu takes up less than 10 percent of the global market and is rarely found on cocktail menus along with other spirits.

In an effort to attract a wider range of consumers, especially in the overseas market, Chinese liquor makers are introducing new technology and new products besides carrying on the ancient tradition of liquor brewing.

In Yibin, the home city of Wuliangye, an academy has been set up to train talents for the industry, while in Luzhou Laojiao Group, liquor-taste masters like Zeng Na are developing cocktails mixed by spirits and fruit juice to *cater* to the lighter taste of beginners.

In recent years, Luzhou Laojiao has been dedicated to rejuvenating the fine traditional Chinese culture and promoting the spread of traditional culture on a global scale.

In 2017, Luzhou Laojiao launched a series of global charity events such as Guojiao 1573, "Let the World Feel the Taste of China" and an international poetry culture conference, and actively participated in the national events to promote the outstanding Chinese traditional culture on the world stage.

"Foreign markets are still wide-open for Chinese liquor producers, and Chinese Baijiu will make a breakthrough in the global market in the next five to ten years," said Lin Feng, general manager of Luzhou Laojiao Group.

New Words and Expressions

1. Bordeaux [bɔːˈdəʊ] *n.* 波尔多（法国西南部港市）
2. landmark [ˈlændˌmɑːk] *n.* 地标
3. subtropical [ˌsʌbˈtrɒpɪkl] *adj.* [地理] 亚热带的
4. sorghum [ˈsɔːɡəm] *n.* 高粱；[作物] 蜀黍
5. unduplicated [ʌnˈdjuːpləˌkeɪtɪd] *adj.* 无复制品的；无法匹敌的
6. sewage-processing plants 污水处理厂
7. infrastructure [ˈɪnfrəˈstrʌktʃə] *n.* 基础设施
8. logistics distribution 物流配送
9. quality supervision 质量监督
10. glutinous [ˈɡluːtɪnəs] *adj.* 粘的；粘性的
11. organic [ɔːˈɡænɪk] *adj.* [有化] 有机的
12. tipple [ˈtɪpl] *n.* 酒；烈酒
13. intangible [ɪnˈtændʒəbl] *adj.* 无形的，触摸不到的
14. cater [ˈkeɪtə] *vi.* 投合，迎合
15. rejuvenate [rɪˈdʒuːvəˌneɪt] *vt.* 使复兴，使恢复活力

Exercises

I. Choose the best answer to each of the following questions.

1. What will contribute to protecting the quality of Baijiu and helping alleviate poverty?
   A. The establishment of the China Golden Triangle of Baijiu.
   B. The local cultivation of high-quality brewing raw materials.
   C. Legendary liquor cellars.
   D. Liquor-making techniques.

2. What advantages are the China Golden Triangle of Baijiu enjoy in Baijiu making?
   A. Increase of population.
   B. Many brewing cellars, Baijiu workshops and tourism destinations.
   C. Subtropical climate, humid weather, clean water, rich soil and basin landscape.
   D. All of the above.

3. What make Luzhou city the special tourism destination for spirits culture experience?
   A. Ancient brewing cellars, workshops and natural liquor caves.
   B. Ideal environment for brewing good Baijiu.
   C. Remarkable inheritor of Baijiu tasting.
   D. A large area of sorghum.

4. What are Chinese liquor makers doing to open the international market?
   A. Introducing new-technology-and-creativity products.
   B. Launching global charity events.
   C. Carrying on the ancient tradition of liquor brewing.
   D. All of the above.

5. Why did the Moutai company put great effort on environmental treatment?
   A. The healthy ecosystem is essential to the quality of Baijiu.
   B. The environmental treatment can help people live longer.

C. The environmental treatment can help creat a special ecosystem.

   D. The healthy ecosystem can make liquor makers more comfortable.

6 What's the function of Baijiu Industry Park in Luzhou?

   A. Inviting in qualified upstream and downstream enterprises increase GDP.

   B. Forming codes of practice and guarantee quality of Baijiu products.

   C. Promoting local agricultural industry, helping farmers to increase their incomes.

   D. Accelerating the development of tourism.

7 What efforts did Wuliangye do to spread the spirits culture to markets around the world?

   A. Developing spirit and fruit juice mixed cocktails to *cater* to the lighter taste of beginners.

   B. Participating in the national events positively.

   C. Introducing new-technology-and-creativity products

   D. Guarantee the quality of products and set up an academy to train talents for the industry.

8 What produced the unique flavor of Moutai?

   A. Century-old brewing cellars and natural wine caves.

   B. Advanced brewing techniques.

   C. Unique bacterial and archaeal diversities and the grain in the region.

   D. Abundant water resources and sorghum fields.

II. Translate the following passage into English.

唐宋时期酒与文艺紧密联系，这可以从唐诗宋词等文艺作品中看出来。这种现象使唐宋成为中国酒文化发展史上的一个特殊时期。明朝是酿酒业大发展的时期，酒的品种、产量都大大超过之前任何一个朝代。到了清代，大酿坊陆续出现，产量逐年增加，销路不断扩大。为扩大和便利销售，各大酿坊统一了酒的品种、规格和包装形式，并开设酒馆或酒庄，经营批发零售业务。

UNIT 5
Manufacture and Classfication of Baijiu

Preview

Wuliangye and Moutai are the representatives of Chinese Baijiu, which smell fragrant and have a high alcohol content, usually more than 30 percent, sometimes reaching 65 percent. Baijiu is typically obtained by natural fermentation. The manufacture of Baijiu began at least before the 2nd century B.C. An historical record mentions that the earliest distillation tools (distillation pan) appeared during the era of the Song Dynasty. *Song-Shi* (*History of Song Dynasty*), authored by Tuo Tuo in 982 A.D., described a method using wheat, barley, sticky rice, etc. to produce a distilled liquor, which is the exact approach for Baijiu production used today.

Section A

Manufacture of Baijiu

Baijiu production technique is a precious national heritage of China, and the present-day Baijiu is the result of technical progress. Traditional Chinese *fermentation* techniques have benefited the Chinese for hundreds of years. Traditional fermented products have improved Chinese dietary habits and enriched the world's food culture. Baijiu is considered to be the backbone of Chinese fermented products since its output has grown rapidly and currently exceeds 12 million tons per year.

Although there are more than 10,000 factories producing Baijiu with their own techniques, the principle of Baijiu production remains the same. In total, there are eight major steps of the production process of Baijiu: (1) ingredient *formulation*; (2) grinding and cooking; (3) mixing and cooling; (4) mixing with *daqu* (yeast for making hard liquor); (5) loading to the fermentation vessel; (6) alcoholic fermentation; (7) *distillation*; and (8) *aging*.

Sorghum has been the primary material of Baijiu. Sorghum is ground to release *starch* with the purpose of increasing the cooking and microbial correlation area and obtaining a desirable cohesion of mass. This step plays an important role in the quality of Baijiu, since soft grinding leads to ineffective *saccharification* and harsh grinding influences the flavor of Baijiu. The aim of cooking is to introduce starch for *gelatinization*. Water hotter than 85°C and other *additives* are mixed in order to obtain a uniform *texture* and desirable flavor. Reduce the temperature by cooling to prepare for the mix with active *microbiota* (daqu). Loading fermented materials (grains with daqu) into an *earthen jar* when their temperature reduces to between 18°C and 20°C. Alcoholic fermentation is typically carried out in an earthen jar. The jar is in cycle use from one *batch* to another. The fermentation time is dependent on various factors such as climate and *moisture* content. Alcoholic fermentation takes about one month.

Distillation is the key step in the development of flavor of Baijiu. The efficiency of distillation depends on the steam flow rate, water content, distillation speed and *porosity* of materials. It has been shown that a lower steam flow rate may not provide the *thermal* condition for the complete *evaporation* of *ethanol*, while a higher steam flow rate may cause the fermented grains to stick together and increase the *diffusion* resistance. Aging plays an essential role in the flavor of Baijiu, since a variety of *aromatic* compounds (mainly *acids and esters)* are balanced during this process through physical changes. In general, the aging time for sauce-flavor Baijiu is more than 3 years, while at least 1 year is required for strong- and light-flavor Baijiu.

High-temperature *stacking fermentation* is the key process in the production of Baijiu. During the daqu-making process, high temperature is required to obtain daqu dominated by *thermophilic* bacteria. However, yeasts and *molds* play significant roles in alcohol fermentation, since they convert fermentable sugar into alcohol and aroma compounds. Due to this, high-temperature stacking fermentation is introduced directly after adding daqu to sorghum. The principle of this technique is rather simple, that is, to stack the fermented material for a few days (2–4 days) before alcoholic fermentation to increase the number of *microorganisms*, particularly yeasts, and balance the chemical components.

Using *mud pit*: A mud pit is a cellar made from mud. The average volume of the mud pit is from 6 m^3 to 8 m^3. When the temperature of the fermented materials reduces to between 20°C to 21°C, the mixture is loaded into the mud pit. Experience has shown that the pit not only provides a place for fermentation but also contributes to the flavor of Baijiu.

Using *back-slopping technique*: Baijiu is obtained by alcoholic fermentation followed by distillation, and the products are normally unstable. Back-slopping technique is introduced by adding the fermentation *residue* at the beginning of alcoholic fermentation for a better regulation of the fermentation process. This technique allows for an *optimal composition* of the *microbial communities*, which in turn increases the success of natural fermentation. This technique is regarded as an important step compared to other types of Baijiu production.

Baijiu produced in southern China, such as Wuliangye and Jiannanchun, primarily use sorghum, rice, glutinous rice, wheat, and corn as the raw materials, while that produced in northern China use sorghum as the only material. Strong-flavor Baijiu produced in southern China use wheat as the raw material to make daqu, while those in northern China use *barley*, wheat, and peas as the raw materials to make daqu. Daqu is normally made into a *block shape* and ground before mixing with sorghum and other grains. Due to the *subtropical monsoon* climate, alcoholic fermentation takes from 60 to 90 days for Baijiu produced in southern China. Compared to southern China, northern China has the characteristics of low moisture and long daylight. Therefore, alcoholic fermentation takes approximately 45 to 60 days.

In summary, each type of Baijiu has its own specific and critical techniques that result in the different flavors of Baijiu. They are all associated with the specific daqu-making process and alcoholic fermentation process. Changes in each step will alter the *microbial profiles*, which consequently alters the *microbial metabolites* as well. As a result, Baijiu with different flavors could be produced.

New Words and Expressions

1. fermentation [ˌfɜːmenˈteɪʃ(ə)n] n. 发酵
2. formulation [ˌfɔːmjʊˈleɪʃn] n. 配制
3. daqu n. 大曲，又称块曲或砖曲，以大麦、小麦、豌豆等为原料，经过粉碎，加水混捏，形似砖块，大小不等，让自然界各种微生物在上面生长而制成，统称大曲
4. distillation [ˌdɪstɪˈleɪʃn] n. 精馏，蒸馏，净化
5. aging [ˈeɪdʒɪŋ] n. 陈化
6. starch [stɑːtʃ] n. 淀粉
7. saccharification [səˌkærɪfɪˈkeɪʃn] n. 糖化（作用）
8. gelatinization [dʒəˌlætɪnaɪˈzeɪʃən] n. 凝胶化（作用）
9. additive [əˈdɪktɪv] n. 添加剂，添加物
10. uniform [ˈjuːnɪfɔːm] adj. 一致的，统一的
11. texture [ˈtekstʃə] n. 质地
12. uniform texture 质地均匀
13. microbiota [maiˈkrəbaiˈəutə] n. [微] 微生物区
14. earthen jar 瓦罐
15. cycle use 循环使用
16. batch [bætʃ] n. 一批；一炉
17. moisture [ˈmɒɪstʃə] n. 水分；湿度
18. porosity [pɔːˈrɒsɪtɪ] n. 孔隙度
19. thermal [ˈθɜːm(ə)l] adj. 热的；热量的
20. evaporation [ɪˌvæpəˈreɪʃn] n. 蒸发；消失
21. ethanol [ˈeθəˌnɒl] n. [有化] 乙醇，[有化] 酒精
22. diffusion [dɪˈfjuːʒ(ə)n] n. 扩散，传播
23. aromatic [ærəˈmætɪk] adj. 芳香的，芬芳的
24. acids and esters 酸类和酯类
25. stacking fermentation 堆积发酵
26. thermophilic [ˌθɜːməʊˈfɪlɪk] adj. 适温的，喜温的
27. mold [məʊld] n. 霉菌

28　microorganism [ˌmaɪkrəʊˈɔːɡənɪzən] *n.* [微] 微生物
29　mud pit 窖池
30　back-slopping technique 续槽技术（通过在酒精发酵开始阶段添加发酵残渣以更好地调节发酵过程）
31　residue [ˈrezɪˌdjuː] *n.* 残渣
32　optimal composition 最佳组成
33　microbial communities 微生物群落
34　barley [ˈbɑːlɪ] *n.* 大麦
35　block shape 块状
36　subtropical monsoon 亚热带季风气候
37　microbial profiles 微生物分布
38　microbial metabolite 微生物代谢产物

Exercises

I. Answer the following questions.

1　What are the major steps of the production process of Baijiu?
2　What is the different consequence made by the lower steam flow rate and higher steam flow rate in the process of distillation?
3　Why aging plays an essential role in the flavor of Baijiu?
4　Why is high-temperature stacking fermentation the key process in the production of Baijiu?
5　What is the function of using mud pit?
6　What is back-slopping technique?
7　What does the fermentation time depend on?
8　How much time does alcoholic fermentation take?

II. Discuss the following questions.

1　Give a brief introduction of distillation and stacking fermentation.
2　Tell some stories about the brewing of liquor in history.

III. Match the items in column A with those in column B.

A	B
1. 发酵剂	____a. uniform texture
2. 固态发酵	____b. yeasts and molds
3. 酱香味白酒	____c. acids and esters
4. 高温堆积发酵	____d. solid state fermentation
5. 窖池	____e. fermentation starters
6. 续糟技术	____f. mud pit
7. 酵母和霉菌	____g. ingredient formulation
8. 酸类和酯类	____h. sauce-flavor Baijiu
9. 配料构成	____i. high-temperature stacking fermentation
10. 质地均匀	____j. back-slopping technique

IV. Fill in the following blanks with the words given below. Change the form when necessary.

aromatic	cycle use	additive	distillation	barley
formulation	diffusion	fermentation	aging	texture

1. Do you know the material is in _____?
2. The product description of food _____ shall not contain exaggerated or false advertising.
3. He tried to obtain another material by _____.
4. Its whiskey is made from _____ grown in the northern Rockies and Rocky Mountain water.
5. The better prepared students have to _____ the concepts, so that they can explain them, which really helps.
6. The invention of printing helped the _____ of information.
7. It is the _____ process that turns grape juice into wine.
8. _____ is the last step of the production process of Baijiu.
9. This cloth has a silky _____.

10　The major things they seemed to be selling were _____ oils in little cans.

V. Translate the following passage into English.

中国人在七千年以前就开始用谷物酿酒。总的来说，不管是古代还是现代，酒都和中国文化息息相关。长久以来，中国的酒文化在人们生活中一直扮演着重要的角色。我们的祖先在写诗时以酒助兴，在宴会中和亲朋好友敬酒。作为一种文化形式，酒文化也是普通百姓生活中不可分割的部分，比如生日宴会、送别晚宴、婚礼庆典等。

VI. Write a passage according to the given outline with no less than 150 words.

A Brief Introduction on Manufacture of Baijiu（简要介绍白酒酿造过程)

1）我国白酒酿造史悠久；
2）白酒酿造主要过程；
3）不同酿造过程产生不同的白酒口味。

Section B

Classification of Baijiu

[A] Hundreds of different types of Baijiu are produced by various processes in different regions of China. They can be distinguished by the manufacturing techniques, *fermentation starters,* and product flavors. Various Baijiu with different flavors are produced due to their different fermentation processes and operation conditions.

By Manufacturing Techniques

[B] SSF (*Solid State Fermentation*) Baijiu. SSF is a process in which microbial cultures are grown on a solid matrix in the absence of a liquid (aqueous) phase. This method has been commonly used for most of the well-known Baijiu production techniques, and it has a long history and has been passed on through several generations. This type of Baijiu is typically produced from  grain such as sorghum, wheat, rice, glutinous rice, and maize by a complex SSF process, which consists of: (1) material preparation; (2) daqu making; (3) SSF; (4) solid-state distillation; and (5) aging. This technique results in fermented materials containing approximately 60% water. Therefore, SSF is regarded as a method that could produce the maximal types of Baijiu, with each product having different flavor and characteristics.

[C] *Semi-SSF* Baijiu. Semi-SSF Baijiu is represented by Guilin sanhua and Quanzhouxiangshan. The fermentation process is operated under semi-solid state.

[D] *Liquid-state fermented* Baijiu. Liquid-state fermented Baijiu is represented by Red Star Erguotou. All the production processes include saccharification, fermentation, and distillation performed under liquid state.

By Fermentation Starters

[E] According to the fermentation starters used, three different types of Baijiu can be distinguished.

[F] Baijiu produced using daqu as the starter is represented by four most famous Baijiu brands: Moutai, Wuliangye, Fenjiu and Luzhou Laojiao. Daqu is made from grain like raw wheat, barley, and/or peas. The wetted materials are transferred to a molding press and shaped as a brick, each weighing approximately 1.5kg to 4.5kg, with either a flat surface or one side in a convex shape. Because of its big size, it is therefore named as daqu (big starter). In general, four categories of microorganisms (bacteria, yeast, *filamentous fungi,* and *actinomycetes*) can be found in daqu. They provide different enzymes and flavor precursors for the production of Baijiu. Baijiu produced with daqu is enriched in flavors, with a long fermentation time and low alcohol production.

[G] Baijiu produced with *xiaoqu* as the starter is represented by Guilin sanhua and Liuyanghe. Compared to daqu, xiaoqu is made from rice or rice bran. Unlike daqu, only very a few types of microorganism are present in xiaoqu, including *Rhizopus, Mucor, lactic acid bacteria,* and yeasts. These microorganisms are primarily good fermentation performers; therefore, a small quantity of the starter and short fermentation period are required in the production of this type of Baijiu. However, due to a few types of microorganism involved, Baijiu produced with xiaoqu contains less flavor than that produced with daqu.

[H] Baijiu produced with *fuqu* as the starter is represented by Erguotou. Fuqu is different from the other two starters and is made from bran and contains only pure culture of *aspergillus*. Aspergillus is a well-known starch degrader and can convert starch into fermentable sugars. Baijiu produced with fuqu has characteristics of light flavor and high liquor production.

By the Flavor of Baijiu

[I] Based on their flavor, Baijiu can be divided into three major and nine minor categories.

[J] *Sauce-flavor Baijiu*, such as Moutai and Langjiu, provides a flavor resembling soy sauce, full-body and a long-lasting aroma. The major representative *aroma* compounds are phenolic compounds, primarily *syringic acid*, along with small quantities of *amino acids,* acids, and esters.

[K] *Strong-flavor Baijiu,* such as Luzhou Laojiao and Wuliangye, has the characteristics of fragrant flavor, soft mouthfeel, and endless aftertaste. The representative aroma compounds are predominantly *ethyl hexanoate* in harmonious balance with *ethyl lactate, ethyl acetate, and ethyl butanoate.*

[L] *Light-flavor Baijiu,* such as Fenjiu and Red Star Erguotou, gives a pure and mild flavor, mellow sweetness and refreshing aftertaste. The major aroma compounds are ethyl acetate in balance with considerable levels of ethyl lactate.

[M] The characteristics of these three flavor types are very typical and representative, and they comprise approximately 60% to 70% of Baijiu in China. The production techniques for these three types of Baijiu are standardized and stereotyped. In addition to these three types, there are other types of Baijiu of specific flavor and aroma produced with their unique techniques such as Baiyunbian, Dongjiu and Sitir. With the development of science and technology and using various starting materials, Baijiu of various flavors can be produced. More strict and accurate standards are required in the future to distinguish Baijiu based on their flavor-enriching compounds.

[N] Other classifications of Baijiu can also be divided into high alcohol content (above 50% v/v), medium alcohol content (41%–50% v/v), and low alcohol content (<40% v/v); and according to the raw materials used for fermentation, there are sorghum-based Baijiu, corn-based Baijiu, and rice-based Baijiu.

New Words and Expressions

1. fermentation starter 发酵器
2. solid-state fermentation (SSF) 固态发酵
3. semi-SSF 半固态发酵
4. liquid-state fermented 液态发酵的
5. filamentous fungi 丝状真菌
6. actinomycetes [ˌæktɪnə(ʊ)ˈmaɪsiːt] [微]放线菌
7. xiaoqu n. 小曲，以稻米为原料制成的糖化剂，与大曲相比，小曲要小很多，一般是球状，如鹌鹑蛋大小，或者更大一些，如鸡蛋一般大
8. rhizopus [ˈraɪzəʊpəs] n. 根霉菌
9. mucor [ˈmjʊkɔ] n. 毛霉菌
10. lactic acid bacteria 乳酸菌
11. fuqu n. 麸曲，采用纯种霉菌菌种制成的糖化剂，以麸皮为原料经人工控制温度和湿度培养而成
12. aspergillus n. [ˌæspəˈdʒɪləs] 曲霉菌
13. sauce-flavor n. 酱香味
14. aroma [əˈrəʊmə] n. 芳香
15. phenolic [fɪˈnɒlɪk] adj. [有化] 酚的
16. phenolic compounds 酚类化合物
17. syringic acid 丁香酸
18. amino acids 氨基酸
19. strong-flavor n. 浓香味
20. ethyl hexanoate 己酸乙酯
21. ethyl lactate 乳酸乙酯
22. ethyl acetate 乙酸乙酯
23. ethyl butanoate 丁酸乙酯
24. light-flavor n. 轻香味

Exercise

I. Reading comprehension.

The following ten statements are derived from the passage. Each statement contains information given in one of the paragraphs. Identify the paragraph from which the information is derived. Each paragraph is marked with a letter.

1. Xiaoqu is different from daqu, which is a small starter and made from rice or rice bran. Unlike daqu, only very a few types of microorganism are present in xiaoqu.
2. The maximal types of Baijiu are produced by the traditional process, which is called solid state fermentation. It has a long history and has been passed on through several generations.
3. Microorganisms are found in daqu, which is made from grain, wheat and barley. The wetted materials are transferred to a molding press and shaped as a brick.
4. There are various process to produce Baijiu in different regions of China and the diverse process can produce different and unique flavors.
5. Fuqu contains lots of aspergillus, which is a starch degrader and can convert starch into fermentable sugars.
6. Luzhou Laojiao and Wuliangye are the well-known representatives of strong-flavor Baijiu, which have the characteristics of fragrant flavor, soft mouthfeel, and endless aftertaste.
7. Based on the alcohol content, Baijiu can be divided into high alcohol content (above 50% v/v), medium alcohol content (41–50% v/v), and low alcohol content (<40% v/v).
8. In the future to judge the flavor of Baijiu, more strict and accurate standards are required to distinguish its flavor compounds.
9. Ethyl acetate and ethyl lactate are the major aroma compounds of light-flavor Baijiu.

10 Syringic acid, together with a small deal of amino acids, acids, and esters form the main aroma compounds of sauce-flavor Baijiu, which tastes like soy sauce.

II. Translate the following passage into English.

我国是"世界稻文化起源地",酿酒原料丰富、充足,酿酒历史源远流长。考古挖掘出的一些酿酒、饮酒的陶器(pottery)证明,早在五千年前中国人就已经能酿造低度的粮食酒了。甲骨文(inscriptions on bones or tortoise shells)和金文的"酒"字,以及丰富多彩的青铜酒器均说明在商周时期造酒业就已经十分发达了。蒸馏器的发明,在中国造酒业发展过程中具有里程碑意义。在距今大约一千多年前的唐代,中国人采取蒸馏法,酿造出了酒精度较高的粮食酒,也就是人们现在所说的白酒。

UNIT 6 Tasting

Preview

Sharing, instead of enjoying alone, is the key to Baijiu tasting which plays a lubricative role in social contact. There are many popular sayings in China about Baijiu, like "You can't set the table without alcohol", which is to say that you cannot make a banquet with the absence of alcohol; and "When one drinks with a friend, a thousand glasses are not enough", which is generally accepted as a way of making friends. Knowing how to appreciate Baijiu, people can easily enjoy the process and be come friends with those who share the same ideas of tasting without too much communication.

Section A

Enjoying Tasting Baijiu

There are various types of Baijiu. With a sufficient preparation before tasting and be trained with systematic method, the tasters will be rewarded by pleasure. Knowing how to taste Baijiu could be a skill to enlarge people's circles, meanwhile, deepen their friendships.

Preparation

PLACE: The evaluation of Baijiu is influenced by the environment so it's of vital importance to choose a perfect place. First, the air should be clean without any smog or odor. Second, the natural light or white light is preferred. The bars, for example, are not suitable for its mixed smell and colorful lights. Third, there is supposed to be clean sink so as to wash the glasses and other tools. Last, the temperature there enjoys 18-20°C because both higher and lower affect diffusion of the aroma.

GLASS: The results come out under the impact of the size, the color, the quality and the capacity of the glasses. Choosing the right glasses is crucial to getting accurate results. Standard Baijiu glasses prefer goblets with the quality of great transparence, smooth and bright finish, proportionate thickness and shape of tulip (the goblets in the given

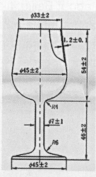

pictures are of exclusive use: hight: 54±2mm, maximum diameter: 45±2mm, upper *caliber*: 33±2mm, total volume: 50-55ml). The goblets should be radically washed and put into a completely smell-free environment.

TASTER: Tasters are supposed to clean their oral *cavity* in advance and avoid strong-flavored food to protect the original flavor of Baijiu. To distinguish the flavor accurately, tasters shall not put on any perfume or flavory cosmetics, let alone smocking. Furthermore, they ought to keep enough sleep and be healthy in order to maintain their sensitivity.

OTHER THINGS: A4 blank papers for observing the color; clean water to clean the tools and mouth; flavor free bread or cookie for recovery of the sensitivity after tasting multiple times; notebooks and pens; aroma free napkin; barrels for spitting and other things personally necessary.

Four Steps to Taste

Once Baijiu is all on the table, the first thing is to observe the volume in each goblet and make sure all the samples are of the same volume (same fluid level). There are certain orders to obey that the Baijiu of shallow color is always prior to that of deep color; that Baijiu of light aroma is always prior to that of strong aroma; that Baijiu of low alcohol concentration is always prior to that of high alcohol concentration. The four steps of tasting are as the follows: observing, smelling, tasting, evaluating.

Observing

Put the sample on a white paper and then look at it both horizontally and down to observe the color of it: colorless or bright yellow. Next, find out whether it is clean: transparent or *turbid*. Take up the goblet and shake it to observe the drops trickling down inside and to see whether there is suspension solid or sediment. Record all the outcomes of above.

Smelling

Smelling is a pivotal approach. Take up the goblet and keep your nose 1-3cm away from the it, smell for the first time. Fan the aroma by one hand for the second and smell closely to the cup for the third. Always remember to *exhale* before gradual *inhaling* and not to exhale to Baijiu. Repeat those approaches for 3 times at most and always record every step accurately. After you smell one sample, take an interval and then go for another. If there are too many samples, smell from the first sample to the last one and do it backward, then repeat the steps. Pick out the most characteristic ones then those of similar characteristics. Amend the records after each step.

To distinguish the aroma, you should find out that whether the aroma is pure; whether there is peculiar smell; whether the aroma is light or strong; what the aroma type of the sample is; how long can the aroma maintain.

Tasting

Remember to taste Baijiu of extremely heavy aroma and of peculiar smell last in case that their aroma mess with that of others. Generally, you should sip only a little at a time (0.5-2ml). When the liquid overspreads your tongue, you can try to distinguish its taste and spit it out or swallow it after 5-10 seconds. The aroma will go through your nostril when you exhale and that is the moment you should decide whether the aroma *pungent* and whether it is strong or light. Use your tongue to feel the balance of the taste and move and lick around the cavity to taste the astringency. Then drink a moderate amount of Baijiu. This time, you need to pay attention to the persistence of the aroma, purity of the aftertaste, also, that whether the aftertaste is astringent or sweet, whether there is *lingering* aroma, whether your throat feel acrid or unpleasant. Remember to take notes while you taste. You shall not taste over three times but need to wash mouth or eat some bread or cookie to keep your mouth clean and avoid *gustatory overstrain*.

For evaluating Baijiu of the same type but different ranking, the quality is represented by their softness, harmony, fullness, refreshment, lingering aroma, aftertaste and so on. Any Baijiu, despite of the types, with bitter aftertaste or tasting of acerbic or peculiar flavor is of low-grade quality.

Criterions for Tasting

Softness: mellow, mild, smooth, harmony, pungent, *poignant*

Fullness: rich, heavy, complex, mellow, *plump*, thin, light, poor

Harmony: harmonious, balanced, delicate, inharmonious, unbalanced

Purity: clean, pure, astringent, discordant

Persistence: lingering, long, short

Alcohol concentration: low: ≤40% middle: 41-50% high: ≥51%

Evaluation

Evaluate Baijiu based on the forementioned three steps, considering their style, quality, price. Then score by percentage.

Classic types

Srtong-aroma style: **Strong-aroma Baijiu is fermented in earthen pits. It is the most popular and widely produced category. It could be distilled from one or more grain. Various and unique geographical and technological factors attributes to the exclusive flavor of Baijiu**

as Wuliangye, Luzhou Laojiao and Swellfun of Sichuan are known for their fruity sweetness. Wuliangye is the most famous representative of this style.

Sauce-aroma style: Sauce-aroma Baijiu derives its name from its distinct and lingering fragrance, which is said to resemble soy sauce. Tasting exquisite with prominent sauce aroma and thick body, whose aftertaste lasts long as well as lingers in the empty glass. There are historical brands such as Moutai (Guizhou province), Langjiu (Sichuan province), Wulinjiu (Hunan province) but also fresh brands like Yongfujiangjiu from Wuliangye (Sichuan province) and classic sauce-aroma Baijiu Tunzhihu series from Shede Spirits (Sichuan province).

Light-aroma style: Light-aroma Baijiu is fermented with daqu made from barley and peas, and typically has a mild floral sweetness, which is typically represented by Fenjiu (Shanxi province) and Erguotou (Beijing). Light-aroma Baijiu gives preeminence to mildness, coolness and softness. It is harmonious and clear.

Rice-aroma style: Represented by Sanhuajiu (Guangxi), rice-aroma Baijiu softens your tongue with its sweetness, coolness and great aftertaste. It is distilled from long-grain rice, glutinous rice or a combination of the two and fermented with xiaoqu. At its best, rice aroma is smooth and mild.

Chi-aroma style: Chi aroma's name refers to *douchi*, a salty Chinese condiment made from fermented beans. Represented by Yubinshaojiu (Shiwan Distillery, Guangdong province), fermented soya flavor clears your mouth with its special soya flavor and smoothy body as well as neat aftertaste.

Feng-aroma style: Named after Xifengjiu (Fengxiang County, Shaanxi Province), Feng-aroma combines aspects of the strong- and light-aroma categories. It is aged in Jiuhai (a large liquor container): giant rattan baskets filled with cloth sacks hardened with vegetable oil, beeswax and pig's blood. This category of Baijiu are notable for their fruity aromas, grainy taste and expanding finishes.

Mixed-aroma style: Mixed-aroma Baijiu is produced by combining production techniques or blends from two different categories, but either of which may play the predominant role. Baiyunbian of Hubei province is the representative of this style with strong aroma in its sauce aroma; and Yuquan of Heilongjiang province with sauce aroma in its strong aroma.

Laobaigan-aroma style: Represented by Laobaigan of Hengshui Ruitian Distillery, Laobaigan-aroma Baijiu is similar to light-aroma Baijiu in almost every aspect but uses wheat instead of barley and peas to make its qu. With less than six months aging, it has a fruity flavor, exquisite body, soft and endless taste overshadowed by a searing alcoholic flame.

Sesame-aroma style: Represented by Jingzhibaigan of Jingzhi Distillery of Shandong province and Meilanchun of Jiangsu province, sesame-aroma Baijiu is characterized by a typical sesame odor companied with exquisite, sweet, harmonious flavor, and clear aftertaste.

Te-aroma Style: Te-aroma is associated with Jiangxi Sitir Distillery. Its fragrance is earthy, but its taste is light with a rich, slightly tart aftertaste.

Herbal-aroma style: Named after its producer Dongjiu Distillery of Guizhou province, also known as Dong aroma, herbal-aroma Baijiu is either too sweet or too sour but smells herbal, exquisite and harmonious.

Extra-strong-aroma style: Represented by Jiuguijiu of Hunan Province, extra-strong-aroma Baijiu has a pungent earthy fragrance and a spicy-sweet taste.

Tasting skills

As gustation and *olfaction* get tired of long-time tasting, the first impression on Baijiu generally is proved more accurate.

Baijiu tasting is of 70% smelling and 30% tasting, which means that the quality of Baijiu appears seven out of ten under aroma. However, it's not saying that strong aroma makes better Baijiu but tells us to evaluate basing on harmony and softness. The quality shall not be inferior with exquisite and natural aroma.

You shall not drink as soon as you finish smelling or drink every sip, for that the moment you swallow the sip, gustation stimulated by that, your olfaction will not be able to taste thence. In case that you forgot any transient feeling or information in the process, you'd better to grasp any sensory actions and write them down.

Tasting Note

Name of Baijiu		Taster		Place		Date	
L O O K I N G	Color	Peculiar color		Slight yellow		Colorless	
	Transparency	Turbid				Transparent	
	Suspended Solid	Yes				No	
	Sediment	Yes				No	
S M E L L I N G	Purity	Peculiar smell				Pure and natural	
	Richness	Thin		Middle		Rich	Heavily Rich
	Aroma Characteristic	Ingredients aroma		Fermentation aroma		Chen aroma	
	Persistence	Short		Long		Lingering	

续表

	Name of Baijiu		Taster			Place			Date		
TASTING	Body		Thin			Middle			Thick		
	Softness		Poignant		Pungent	Harmony		Smooth	Mild		Mellow
	Fullness		Poor		Thin	Mellow		Plump	Heavy		Rich
	Harmony		Unbalanced		Rough	Delicate		Coordinated	Balanced		Harmonious
	Purity		Discordant			Astringent			Clear		
	Persistence		Short			Lingering			Long		
	Alcohol Concentration (%)		Low：≤40%			Middle：41-50%			High：≥51%		
EVALUATING	Types	Strong Aroma	Sauce aroma	Light aroma	Rice aroma	Chi aroma	Feng aroma	Mixed aroma	Laobaigan aroma	Sesame aroma	Tearoma	Herbal aroma	Extra-Strong aroma
	Quality		Bad	Mediocre		Ordinary		Middle	Great		Perfect
	Price		Low			Fair			High		
Score(%)			≤ 60	60-70		70-80		80-90	90-100		

New Words and Expressions

1 caliber [ˈkælɪbə] *n.* 口径
2 cavity [ˈkævəti] *n.* 腔
3 lubricative [ˈluːbrɪˌkeɪtɪv] *adj.* 润滑的；有润滑作用的
4 turbid [ˈtɜːbɪd] *adj.* 浑浊的
5 exhale [eksˈheɪl] *vt.* 呼气
6 inhale [ɪnˈheɪl] *vt.* 吸入；猛吃猛喝 *vi.* 吸气
7 linger [ˈlɪŋgə] *vi.* 徘徊
8 pungent [ˈpʌndʒənt] *adj.* 辛辣的；刺激性的
9 gustatory [gʌˈsteɪt(ə)rɪ] *adj.* 味的，味觉的
10 overstrain [əʊvəˈstreɪn] *n.* 过度紧张
11 poignant [ˈpɔɪnənt] *adj.* 尖锐的

12 plump [plʌmp] *adj.* 丰满的
13 strong-aroma style 浓香型
14 sauce-aroma style 酱香型
15 light-aroma style 清香型
16 rice-aroma style 米香型
17 Chi-aroma style 豉香型
18 Feng-aroma style 凤香型
19 mixed-aroma style 兼香型
20 Laobaigan-aroma style 老白干香型
21 sesame-aroma style 芝香型
22 Te-aroma style 特香型
23 herbal-aroma style 药香型
24 extra-strong aroma style 馥郁香型
25 olfaction [ɒlˈfækʃən] *n.* 嗅觉

Exercises

I. Answer the following questions.

1 How do people choose a good place for tasting?
2 What is the standard Baijiu glass like?
3 Make a list of things you should prepare before tasting.
4 What are the major steps of Baijiu tasting?
5 What should we do to distinguish the aroma by observing?
6 What are the criteria to evaluate Baijiu by tasting?
7 What are the major types of Baijiu aroma?
8 What are the characteristics of strong-aroma Baijiu?

II. Discuss the following questions.

1 Talk about your own opinion on the systematic tasting methods.
2 Talk about one of your most impressive experiences of drinking Baijiu.

III. Match the items in column A with those in column B.

A	B
1. 持久度	____ a. seas of alcohol
2. 回味悠长	____ b. balanced and harmonious
3. 酒海	____ c. transparent
4. 清香纯正，醇甜柔和	____ d. cellar aroma
5. 郁金香花型酒杯	____ e. a mild floral sweetness
6. 平衡谐调	____ f. persistence
7. 欠净	____ g. sesame-aroma style
8. 清亮透明	____ h. long aftertaste
9. 窖香	____ i. astringent
10. 芝香型	____ j. a glass of shape of tulip

IV. Fill in the following blanks with the words given below. Change the form when necessary.

| coordinate | mellow | mediocre | turbid | sensitivity |
| sediment | nostril | poignant | overstrain | aftertaste |

1 Most muscle and spinal injuries are from _____.
2 Start with alternate _____ breathing.
3 President Obama wants all countries in the Group of Twenty to _____ their separate efforts to strengthen their economies.
4 It turns out that the patients actually did regain some of their _____ to light.
5 For a(n) _____ and polluted river with highly variable quality, a complex, flexible treatment system should be employed.
6 In other words, instead of regulating the _____ directly, EPA decided that water was a pollutant.
7 San Diego is very quiet, _____, just kind of go to the beach and hang out.

8 The time is long, the pay is _____, nobody respects your contribution, yet you freely choose to work here.
9 The politically divisive impact of their foreign policies could leave a lasting _____.
10 It made the blues even more mournful, even more _____ and cathartic for anyone listening.

V. Translate the following passage into English.

Though often discussed as if it were a single type of spirit, Baijiu is a tree with many branches. In terms of ingredients, production methods and flavors, Baijiu can be worlds apart. Baijiu is divided into four primary categories by smell—strong, light, sauce, and rice aroma—and a handful of smaller, often brand-specific categories. The current classification system is a modern innovation with its limitations, but it is a useful reference for any tasters. Though Baijiu styles have historically been linked to specific regions, geographic lines have began to blur in recent decades and new flavors are continuing to emerge.

VI. Write a passage according to the given outline in no less than 150 words.

Write a comment on the Baijiu you have tasted according to the tasting methods in the text.

Section B

Aroma, the Soul of Baijiu

It's said that "Baijiu with great aroma need no peddling". Great aroma fosters the success of Baijiu and defines the quality of Baijiu. The most important part of tasting is smelling. When it comes to smelling and tasting, tasters take 70 percent for smelling and 30 percent for tasting when they estimate a brand. Experienced tasters can easily grasp the basic information in their hands once they smell.

The aroma of Baijiu is classified into 12 types according to their different characteristics. Different factories may produce the same type but with different aroma characteristic due to the difference of their ingredients, fermenting procedure, blending proportion and storage. For example, in strong-aroma Baijiu, there may be one extremely prominent aroma such as cellar aroma, rich Chen aroma, heavy fragrant aroma. Sometimes, some of them smell refreshing as a modest *orchid* while the others spread their odor over around. That is why aroma is the identification of different Baijiu.

Baijiu of good quality has rich but harmonious aroma which please the taster and last long while those of inferior quality displease people by certain strong but quick-disappearing smell.

There are in total three characteristics of Baijiu aroma:

The first layer: the original aroma of the ingredients which is a key factor in fermentation. Ingredients and *auxiliaries*, grain, yeast and pit mud, for example, will transfer their original aroma to Baijiu during the fermentation.

The prescriptive words of the first layer: grain aroma, rice aroma, peas aroma, herbal aroma, rice bran aroma, qu aroma.

Grain aroma—aroma of simple grain (*danliang*) and mixed grain (*zaliang*). Wuliangye is made of wheat, rice, corn, sorghum, glutinous rice, which are distilled and burnt out their aroma. Baijiu made in this way especially smell of grain. *Oenophiles* think that ethanol comes from sorghum, as sweetness from rice, as richness in glutinous rice, as roughness in corn, and as fragrance from wheat.

Sorghum aroma—sorghum is regarded as the best ingredient to make Baijiu that any other single material cannot match it.

Herbal aroma—Herbal aroma describes its characteristic best in Dongjiu which combines daqu and xiaoqu companied with Chinese medicines. Those Baijiu are prominent for its strong vinegar and herbal flavor.

Qu aroma—Qu aroma is very special that it represents the complete daqu aroma generated from middle and high temperature condition, the main elements to maintain the left aroma in empty glass after drinking. It can be seen in all Sichuan famous strong-aroma Baijiu, which makes Sichuan strong-aroma Baijiu distinct from other provinces'.

The second layer: the aroma is brought from fermentation. Fermentation is an essential step to make Baijiu. In all high-quality Baijiu making, fermentation is strictly and accurately controlled, for that fermented aroma results from *microbic* reaction. As to the principle, compound of *aldehyde* and *ketone* is the *metabolite* of microbes, and the esters reacted chemically in it are the reason that aroma generates, esters of less *molecule* generating strong and special aroma as that of more molecule generating light and mild aroma.

The prescriptive words of the second layer: ethanol aroma, mild aroma, cellar aroma, sauce aroma, rice aroma, baked aroma, sesame aroma, distilled grain aroma, fruity aroma, flower aroma, honey aroma, grassy aroma, nutty aroma, woody aroma, sweet aroma, sour aroma.

Ethanol aroma—rich, soft and harmonious fragrance. It's very typical of Baijiu stored for quite a long time.

Mild aroma—refreshing, light, clean and pure.

Cellar aroma—coming from the aroma of pit.

Sauce aroma—smelling like sauce products.

Baked aroma—smelling being baked and roasted.

Sesame aroma—foundational sesame odor with rich, mild and sauce aroma is thought to be the greatest trait of sesame-aroma Baijiu.

Distilled bran aroma—classical solid fermentation which only generates after distilling maternal fermented grain. The aroma smells of baked and burnt, appearing to be natural.

Honey aroma—Feng-aroma Baijiu stored is mostly in a big container called Jiuhai (seas of alcohol) painted with beeswax and vegetable oil that will melt in Baijiu, generating a peculiar sweet odor and leading to improvement.

The third layer: Aroma comes from years' storage. This aroma is brought about during the long-period storage when all the material and auxiliaries transform on account of inner *transmutation* or chemical interactions.

The prescriptive words of the third layer: chen aroma, oily aroma, jiuhai aroma, jujube aroma

Chen aroma—cellar chen (pit odor or pit mud odor), old chen (a typical aroma of old Baijiu), sauce chen (similar to sauce odor, rich but rough), oily chen (a combination of oily odor and Chen odor) and ethanol chen (refreshing crusted Baijiu aroma after long-time storage, that is hard to form).

Jiuhai aroma—the typical aroma of XiFengjiu stored in the container Jiuhai. Jiuhai aroma appears under the condition that Jiuhai are painted with proportionate blood, egg white, *beeswax* and vegetable oil and dried repeatedly.

Each type of Baijiu has its typical aroma: strong-aroma Baijiu smells of typical grain aroma and cellar aroma while sauce-aroma Baijiu has peculiar sauce aroma and baked aroma. And alcohol aroma and fruit aroma are particular in light-aroma Baijiu, the same as rich and sweet aroma in fermented Feng's, as oily aroma in fermented chi-aroma's, sauce in Laobaigan's, fermented glutinous rice aroma in rice-aroma's, burnt aroma in Te's, sesame aroma and burnt aroma in sesame-aroma's and Chinese medicine aroma in herbal aroma's.

We explore the aroma mentioned above deeper when smelling to search out the clue and basis of evaluating Baijiu. To become a qualified taster, one should have a sensitive nose. Aroma to Baijiu is what olfaction to a taster. Professional tasters need to carefully protect and improve their olfaction, and avoid substances affecting their olfactory sensitivity such as cigarette, obsessive alcohol, some medicine, etc. It is helpful for improving olfactory ability to eat more food with zinc and to exercise often. There must be systematic training of aroma smelling. When trained to taste, the trainees ought to try Baijiu of different aroma, at the same time, try to distinguish different aroma, meanwhile, and try to keep their eyes on daily material which can be used to practice. The more aroma characteristics in their mind, the easier it is to distinguish while tasting, and the evaluation being more accurate. Always remember to summarize and classify Baijiu you have tasted and make notes.

New Words and Expressions

1. orchid [ˈɔːkɪd] n. 兰花
2. auxiliary [ɔːgˈzɪljərɪ] n. 辅助物
3. oenophile [ˈiːnəfaɪl] n. 酒品尝家；酒的行家；嗜酒的人
4. microbic [maɪˈkrəʊbɪk] adj. 微生物的；细菌的
5. aldehyde [ˈældɪhaɪd] n. 醛；乙醛
6. ketone [ˈkiːtəʊ] n. 酮
7. metabolite [mɪˈtæbəlaɪt] n. [生化] 代谢物
8. molecule [ˈmɒlɪkjuːl] n. [化学] 分子；微小颗粒，微粒
9. transmutation [ˌtrænzmjuːˈteɪʃən] n. 变形；变化；演变
10. beeswax [ˈbiːzˈwæks] n. 蜂蜡

Exercises

I. Choose the best answer to each of the following questions.

1. Which of the following is the most important Baijiu tasting method?
   A. Observing.   B. Smelling.   C. Tasting.   D. Evaluating.
2. Where does the second layer of aroma come from?
   A. Material.   B. Additives.
   C. Years' storage.   D. Fermentation.
3. Which one does NOT belong to the first layer of aroma?
   A. Jiuhai aroma.   B. Qu aroma.
   C. Herbal aroma.   D. Bran aroma.
4. Which description fits the cellar aroma?
   A. Complete aroma comes from years' storage.
   B. Aroma results from microbic reaction in pit mud.
   C. Aroma generates after distilling maternal fermented grain.
   D. Complete aroma generates from middle and high temperature condition.

5 Which of the following is the classic aroma of strong-aroma Baijiu?
   A. Fruity aroma.   B. Baked aroma.
   C. Cellar aroma.   D. Honey aroma.
6 What type of aroma does Wuliangye present?
   A. Mild aroma.   B. Sauce aroma.
   C. Strong aroma.   D. Rice aroma.
7 Which of the following is in favor of improving tasters' olfactory sensitivity?
   A. Cigarette.   B. Alcohol.
   C. Cold medicine.   D. Food with zinc.
8 Which one may NOT be a description of Baijiu?
   A. Harmonious.   B. Having pleasurable aroma.
   C. Long-lasting aftertaste. D. Having strong aroma.

II. Translate the following passage into English.

　　一名专业的品酒师需要具备较高的感官敏锐度，但是这种敏锐度不是完全天生的，后天的学习和系统性的训练非常重要。在训练中加强大脑对各种香气和味道的识别能力和记忆能力，不断提高感官的敏锐度和适应性。日常生活中很多食材都可以用来做感官敏锐度训练，如蜂蜜、油、酱油、花香、饭香、芝麻、酒糟等等。只有这样，通过不断的反复训练才能成为一名合格的品酒师。

UNIT 7

Drinking Etiquette

Preview

Baijiu is densely scattered in almost all ancient Chinese documentation. Baijiu plays an indispensable role in rituals and ceremonies. Drinking is an etiquette issue rather than a diet issue. Drinking etiquette emerged simultaneously with Baijiu-brewing, which has rich connotations and becomes a vital component of the catering culture.

Section A

Drinking Etiquette

Eight activities are highly valued in China: music, chess, *calligraphy*, painting, poetry, drinking, flower and tea. Drinking is taken as a *ritual*, a culture, a worldly attitude, and unworldly wisdom. Chinese people drink in three ways: drinking alone, drinking in pairs and drinking in groups. Chinese people regard mood as important, which is expressed with the help of drinking. That is the reason why the intention of the *drunkard* lies not on Baijiu, but on other purposes. Baijiu is not Baijiu; Baijiu is mood. Li Bai[1], showed his refinement in "Among the flowers from a pot of Baijiu, I drink alone beneath the bright moonshine". Han Yi[2] expressed his misery in "When true hearts lie *withered* and fond ones are flown, I drink alone". One traditional saying runs, "If you drink with a *bosom* friend, a thousand cups are not enough; If you argue with someone, half a sentence is too much." Bai Juyi[3], thirsty for drinking with his friend and sharing their own stories, made an invitation, "My new brew gives green glow; my red clay stove flamed up. At dusk it threatens snow. Won't you come for a cup?" Su Shi[4] drank *candidly* and generously with his friends, "Many thousand times I accompanied you, drinking and laughing; we never say goodbye." The drinking way Chinese people prefer is to share with friends and drink *recklessly*. Yan Shu[5] described such a *spectacle*, "Surrounded by distinguished guests, immersed in fragrance of *osmanthus* in floor vases, we *indulge* ourselves in drinking all night long." Flowers and drinking fascinated Li Bai and his friends, with which they entertained themselves, "Where to find new pleasures now that we have enjoyed the flowers? let's go to a pub and kill time happily."

A Dream of Red Mansions[6] depicts Chinese drinking etiquette vividly and *elaborately* on the basis of drinking. As Zhen Shiyin[7] and Jia Yucun[8]'s *binge* drinking in Chapter 1: "At first they sipped slowly, but their spirits rose as they talked and they began to drink more recklessly... The two men became very merry and drained cup after cup." Drink-urging plays an indispensable role in Chinese drinking etiquette. The host constantly urges guests to drink at a banquet, which shows his hospitality and creates a drinking atmosphere. As the birthday party of Wang Xifeng[9] in Chapter 44: "So we have no face, is that it?" protested Yuanyang[10]. "Why, even the mistress *condescends* to drink with us. You usually show us more consideration, but now in front of all these people you're putting on the airs of a mistress. Well, it's my fault for coming. If you won't drink, we'll leave you." Urging Wang Xifeng to drink more, Yuanyang's warmness is fully displayed and their relationship is greatly upgraded. In this way can the host and guests brisk up with *hilarity* at the banquet. Chinese people tend to heat Baijiu while they are drinking, especially in cold weather. As Aunt Xue[11] persuading Baoyu[12] not to have cold drinks in Chapter 8: "Baijiu has an exceptionally fiery nature, and therefore must be drunk warm in order to be quickly digested. If it is drunk cold, it *congeals* inside the body and harms it by absorbing heat from the internal organs."

Drinking adds to fun in a whole year as well as in a whole life. By drinking, the course of life *parallels* four seasons; by drinking at *sequential* festivals, soft and elegant "Chinese style" has survived. Consider drinking on New Year's Eve. New Year's Eve, also known as *Chuxi*, is the last night of the lunar year when people would often stay up late or even the whole night to share their happiness and sorrow in the past year and draw a blueprint for the coming year. It is a time for family reunion, for which the Chinese prepare the most *sumptuous* feast in the whole year. Eating and drinking together, the joy in the family is overflowing. The Qingming Festival falls on the 5th day of the 4th solar month. Chinese people honor nature and the ancestors, which can be traced to ancient times, and still matters a lot in modern time. Cleaning the tombs and paying respect to the dead with offerings are the two important practices to remember late relatives. The offerings include the *deceased*'s favorite food, Baijiu, chopsticks and paper money in the hope that they will enjoy life in the afterlife. Not only is Qingming a period for *commemorating* the dead. It is also time for people to go out and enjoy nature. The grief for the loss is *appeased* and removed temporarily by drinking and the beauty of nature. The Dragon Boat Festival, which comes on May 5th on the lunar calendar, is a traditional festival to the memory of the ancient poet Qu Yuan, who committed *suicide* out of love for his country. People drink *calamus* liquor and *realgar* liquor for *detoxification*, protection against evil and *fending* off something *ominous*. Toad liquor for vigorousness and longevity, together with *albizzia* liquor for *sedation* and peaceful sleep, is popular as well. Mid-Autumn Festival takes place on the 15th day of the 8th month on the Chinese calendar, when the family get together and appreciate the moon. In Chinese beliefs, the full moon is a symbol of a family reunion. Chinese people like osmanthus liquor all the better on that day. It is recorded in *Anecdotes in Tianbao years* that Emperor Xuanzong of Tang Dynasty held a romantic feast on Mid-Autumn night in the palace, all candles put out and music *permeating* the air. They drank in the moonlight. The 9th day of the 9th lunar month is the traditional Chongyang Festival, or the Double Ninth Festival. It is also a tradition that people go hiking on a hill, enjoy *chrysanthemum* and drink chrysanthemum liquor. The 9th lunar month, with clear autumn skies and bracing air, is good time for sightseeing. The chrysanthemum liquor, infused with cornel fruit, has wholesome effects on *assuagement* of pain and regulation of the flow of vital energy. These customs have been passed down till now.

Background Information:

1 Li Bai, a famous Chinese poet of the Tang Dynasty. He has been acclaimed from his own day to the present as a genius and a romantic figure.
2 Han Yi, a Chinese poet of the Yuan Dynasty.
3 Bai Juyi, a renowned Chinese poet and government official of the Tang Dynasty.
4 Su Shi, a Chinese writer, poet, painter, calligrapher, gastronome, and a statesman of the Song Dynasty.
5 Yan Shu, a Chinese statesman, poet, calligrapher and a literary figure of the Song Dynasty.
6 *A Dream of Red Mansions*, composed by Cao Xueqin, is one of China's Four Great Classical Novels. It was written sometime in the middle of the 18th century during the Qing Dynasty. Long considered a masterpiece of Chinese literature, the novel is generally acknowledged to be the pinnacle of Chinese fiction.
7 Zhen Shiyin, a character in *A Dream of Red Mansions*.
8 Jia Yucun, a character in *A Dream of Red Mansions*.
9 Wang Xifeng, one of the principal characters in *A Dream of Red Mansions*. She comes from one of the Four Great Families, and she is known for her great beauty, her wit and intelligence, her vivacious manner, her multiple-faced personality and her fierce sense of fidelity.
10 Yuanyang, the favorite maidservant of Baoyu's grandmother.
11 Aunt Xue, a character in *A Dream of Red Mansions*. She is Baoyu's aunt, and the future mother-in-law.
12 Baoyu, the principal character in *A Dream of Red Mansions*. He is portrayed as having little interest in learning the Confucian classics, much to the despair of his father, Jia Zheng. He would rather spend his time reading or writing poetry and playing with his female relations.

New Words and Expressions

1. calligraphy [kəˈlɪɡrəfɪ] *n.* 书法
2. ritual [ˈrɪtʃʊəl] *n.* 仪式；惯例；礼制
3. drunkard [ˈdrʌŋkəd] *n.* 酒鬼，醉汉
4. wither [ˈwɪðə] *vt.* 使凋谢；使衰弱
5. bosom [ˈbʊz(ə)m] *adj.* 知心的；亲密的
6. candidly [ˈkændɪdli] *adv.* 直率而诚恳地
7. recklessly [ˈrekləsli] *adv.* 鲁莽地；不顾一切地
8. spectacle [ˈspektəkl] *n.* 景象，场面，景色
9. osmanthus [ɒzˈmænθəs] *n.* 桂花
10. indulge [ɪnˈdʌldʒ] *vt.* 满足；使沉迷于
11. elaborately [ɪˈlæbərətli] *adv.* 精巧地
12. binge [bɪndʒ] *n.* 狂欢，狂闹；放纵
13. condescend [ˌkɒndɪˈsend] *vi.* 屈尊；俯就
14. hilarity [hɪˈlærətɪ] *n.* 欢喜；高兴
15. congeal [kənˈdʒiːl] *vi.* 凝结
16. parallel [ˈpærəlel] *vt.* 使……与……平行
17. sequential [sɪˈkwenʃəl] *adj.* 有顺序的
18. sumptuous [ˈsʌmptjʊəs] *adj.* 华丽的，豪华的
19. deceased [dɪˈsiːst] *adj.* 已故的
20. commemorate [kəˈmeməreɪt] *vt.* 庆祝，纪念
21. appease [əˈpiːz] *vt.* 使平息；使和缓
22. suicide [ˈsjʊɪsaɪd] *n.* 自杀
23. calamus [ˈkæləməs] *n.* 菖蒲
24. realgar [rɪˈælɡə] *n.* 雄黄
25. detoxification [diːˌtɒksɪfɪˈkeɪʃən] *n.* 解毒
26. fend [fend] *vt.* 保护；挡开
27. ominous [ˈɒmɪnəs] *adj.* 不吉利的
28. toad [təʊd] *n.* 蟾蜍
29. albizzia [ælˈbɪzɪə] *n.* 合欢

30 sedation [sɪˈdeɪʃən] *n.* 镇静
31 permeate [ˈpɜːmɪeɪt] *vt.* 弥漫
32 chrysanthemum [krɪˈsænθəməm] *n.* 菊花
33 assuagement [əˈsweɪdʒmənt] *n.* 缓和

Exercises

I. Answer the following questions.

1. What activities are highly valued in China?
2. In what ways do Chinese people drink?
3. What does "Baijiu is not Baijiu; Baijiu is mood" mean?
4. What do Chinese people usually express while drinking alone?
5. How did Yan Shu describe the drinking with his friends?
6. What does the host do when he wants his guests to drink more?
7. What do people usually do on the Qingming Festival?
8. What liquors do people drink on the Dragon Boat Festival?

II. Discuss the following questions.

1. In what situation will you drink alone?
2. How do you urge people to drink at the banquet?

III. Match the items in column A with those in column B.

A	B
1. 独饮	____ a. chrysanthemum liquor
2. 对饮	____ b. urge someone to drink
3. 群饮	____ c. drink alone
4. 酒文化	____ d. calamus liquor
5. 劝酒	____ e. drink in pair
6. 畅饮	____ f. drinking culture
7. 雄黄酒	____ g. realgar liquor

8. 菖蒲酒	____ h. osmanthus liquor
9. 桂花酒	____ i. drink in group
10. 菊花酒	____ j. drink recklessly

IV. Fill in the following blanks with the words given below. Change the form when necessary.

| recklessly | spectacle | ritual | wither | hilarity |
| parallel | suicide | ominous | indulge | commemorate |

1. It is annual holiday in France to _____ the fall of Bastille.
2. He drove so _____ that he collided with a tree and landed up in hospital.
3. The grass _____ and died for lack of water under the hot sun.
4. If one thing _____ another, they happen at the same time or are similar, and seem to be related.
5. The scientists found that personal involvement in a(n) _____ is necessary.
6. She was fascinated at the _____ of a rocket launching.
7. Another _____ scene in a nightmare occurred in this small county.
8. The result on the Buckingham Palace balcony was undoubtedly one of the general _____ to all but the Queen.
9. BBC News says the explosion was the work of a(n) _____ bomber.
10. For a few days people can pretend they are free of responsibilities and can _____ themselves.

V. Translate the following passage into English.

In general, Chinese people's hospitality is fully displayed during banquets where interpersonal relations are greatly enhanced through drinking. At the banquet, the host always urges the guests to drink more as a way to express his hospitality. The more the guests drink, the more pleased the host is, as it proves that the guests think highly of the host. If a guest refused to drink, the host would be very disappointed.

Polite toast: This embodies the traditional drinking ethics, i.e., the host politely invites the guests to drink.

Reciprocate toast: This is the return toast by the guest to the host.

Mutual toast: This is the toast between guests. In order to urge others to drink, the toast initiator usually gives all kinds of reasons why the person should drink. If the other person could not refute the reason, he has to drink.

VI. Write a short passage according to the given outline in no less than 150 words.

Outline: 1) Moderate drinking can be good for you;
          2) Drinking will do harm to an empty stomach;
          3) Don't force others to drink.

Section B

Drinking Etiquette of Ethnic Minorities

China is a multi-ethnic country. Diverse drinking etiquette is found in the ethnic groups.

Mongolians are good at singing, dancing and drinking. On grand ceremonies, the local leader or the *prestigious* man lifts the cup, dips his ring finger of right hand into *Koumiss* and *flicks* it to the sky, which shows respect to the god in the sky. Then he repeats the action and flicks to the earth, which expresses respect to the god on the earth. He does it for a third time and flicks to the west, which pays a *tribute* to ancestors. For Mongolians, Koumiss is a sacred beverage, which plays an indispensable role when the host entertains his guests. It is the most popular and the least polite to serve guests with Koumiss. The host pours Koumiss into a cup or silver bowl and holds it on *Hada*, a piece of silk as a greeting gift. The host serves three bowls in a row. His thanks, along with the first bowl, go to God for the sunshine; to the earth for the fortune with the second; to the world for happiness with the third. It is commonly seen that the host offers Baijiu to you respectfully, which forces you to have the drink. The host presents the second bowl to you immediately after you *gulp* down the first one. You try to decline it. On this occasion, the host smiles at you and sings a song to you. He raises the bowl over his head while singing. He does so much for you that you have to take the bowl. You must drink up once you take the bowl, which is the custom. The host offers the third bowl with both hands, singing a toasting song with one leg down. The host would kneel down on the ground if you don't take it. It is hard for you to turn down such kindness. When the fourth bowl is presented to the guest after three continuous drinks, he blesses the host or sings a *paean* in a loud voice, one hand holding the bowl, the other hand up. Mongolians show respect to guests by getting them drunk. When the guests are as drunk as *skunks*, they have earned great respect from the host. The feast never ends until the guests get drunk. The host will be pleased when guests get drunk, which shows they look up to him; The host will be frustrated when guests say no to the toast, which shows they look down upon him. For Mongolians, Koumiss is the cream of beverage, with which proposing a toast shows hospitality and respect to guests. Mongolians urge guests to drink by singing instead of dishes. Singers choose proper songs for each situation, which sound sweet and *melodious*.

Yi people are very hospital and fond of drinking. All people regardless of age or gender are quite good drinkers. Yi people have their own drinking proverbs. "Tea is for Han people while Baijiu is for Yi people." "Why do you come to the world as a man if you don't drink?" "Feast is nothing but Baijiu." So Baijiu matters most. Dishes are not must-haves while drinking. Yi people have a unique custom of drinking, "circle drinking". Friends or strangers sit in a circle on the ground on any occasion or at any place. Without dishes, they fill a big bowl with Baijiu and drink in turn while chatting. One passes

it to another after one takes a sip. People drink from the bowl one by one, which is "circle drinking". Baijiu is absolutely necessary in Yi's life, which is a culture, a tradition and a symbol. Drinking is not only a custom, but also a means of making a resolution or an *oath*. Yi people tend to have blood drinks when they make a decision on a proposal. Killing a roaster on the spot and dropping its blood into the Baijiu, the two sides or the people concerned lift the bowls and give a *pledge*. Afterwards, they empty the bowls, which means it is a deal. On all festivals and holidays, Yi families carry their own Baijiu jars, into which one or several straws reach. Men and women, old and young, sitting in a circle in small groups, suck up Baijiu from the straws. They cheerfully talk a lot about what is going on in the world, and how the interpersonal relationship works while drinking. Friendship, complete with Baijiu, livens things up. This is known as "Zajiu".

Drinking casts a spell on people, which converts strangers into friends and acquaintances into bosom friends. Drinking helps to enhance the relationship between relatives and sometimes functions as a match maker. Due to the strong ethnic flavor, "Zajiu" has been popular with Yi people.

Zhuang people at Tiandong county in Guangxi Zhuang Autonomous Region have a drinking custom called "spoon-come and spoon-go", which is an exclusive means of drinking in the world. They are not used to drinking directly from the big bowl or the cup. They tend to use spoons. They pour Baijiu into a big bowl before drinking, and propose toasts to each other. The host spoons up Baijiu and delivers it to the guest's mouth. The guest must take the drink. The host will be disappointed if the guest refuses. The guest has to give a spoon of Baijiu in return after he drinks the toast. The guest picks up another spoon, spoons up Baijiu from the same bowl and "pours" it into the host's mouth. No matter how long the drinking lasts, spoons go back and forth, which is the "spoon-come and spoon-go."

Dong people are hospitable too. Like other ethnic minorities, they take Baijiu as a present and entertain guests with it. Dong people stop guests with road-block drinks at the gate of a village, at the entrance of a lane or a street, under the drum tower, at the end of a bridge, or at the door of their house. Lads and girls in full dress stand side by side in a row. The guests have to sing back to the girls and drink the big bowls of rice liquor, which is brewed in Dong's households. If the guests don't, it is impossible for them to get through.

New Words and Expressions

1. Mongolian [mɒŋˈgəʊlɪən] *n.* 蒙古族人
2. prestigious [preˈstɪdʒəs] *adj.* 有名望的；享有声望的
3. Koumiss [ˈkuːmɪs] *n.* 乳酒；马奶酒
4. flick [flɪk] *vt.* 轻弹
5. tribute [ˈtrɪbjuːt] *n.*（尤指对死者的）致敬
6. gulp [gʌlp] *vt.* 大口地吸
7. paean [ˈpiːən] *n.* 赞美歌，欢乐歌
8. skunk [skʌŋk] *n.* 臭鼬
9. melodious [məˈləʊdɪəs] *adj.* 悦耳的；旋律优美的
10. oath [əʊθ] *n.* 誓言，誓约
11. pledge [pledʒ] *n.* 保证，誓言

I. Choose the best answer to each of the following questions.

1. Who is the first one to lift the cup on Mongolian grand ceremonies?
   A. The host. B. The prestigious man.
   C. The guest. D. The tourist.

2. What does it mean when a Mongolian flicks the liquor to the west?
   A. He expresses respect to the god in the sky.
   B. He looks up to the god on the earth.
   C. He pays a tribute to ancestors.
   D. He shows his loyalty to his tribe.

3. What is the sacred beverage for Mongolians?
   A. Highland barley. B. Koumiss.
   C. Wine. D. Bile liquor.

4 What is the guest expected to do when he is offered the fourth bowl of liquor?

   A. To kneel down on the ground.

   B. To hold the bowl over his head.

   C. To directly refuse the bowl.

   D. To bless the host or sing a paean in a loud voice.

5 What do Yi people tend to do when they make a resolution?

   A. Drink in a circle.   B. Drink Zajiu.

   C. Have blood drinks.   D. Have Koumiss.

6 What do Zhuang people usually use while drinking?

   A. A spoon.   B. A cup.   C. A bowl.   D. A jar.

7 How many dishes are usually served when Yi people drink Zajiu?

   A. No dishes.

   B. Four dishes.

   C. As many dishes as possible.

   D. Dishes as the friends ask for.

8 Where do Dong people NOT set up roadblocks?

   A. At the entrance of a lane.

   B. At the end of a bridge.

   C. In a house.

   D. At the gate of a village.

II. Translate the following passage into English.

中国人有婚丧嫁娶、生日庆典、庆功祭奠、奉迎宾客等民俗活动，酒都在其中扮演着重要角色。农事节庆时的祭拜庆典若无酒，缅怀先祖、追求丰收的情感则无法寄托；婚嫁之无酒，白头偕老、忠贞不二的爱情无以明誓；丧葬之无酒，后人忠孝之心无以表述；生宴之无酒，人生礼趣无以显示；践行洗尘无酒，壮士一去不复返的悲壮情怀无以倾述。总之，无酒不成俗；离开了酒，民俗活动便无所依托。

UNIT 8

The Selection of Famous Baijiu

Preview

According to a series of classification methods, there are many distinctive varieties of Baijiu. Although it is classified by production techniques, various ingredients and other regional variations, it's worth mentioning that Baijiu can be officially categorized by its own flavor, which termed as "aroma". Many Baijiu brands are available in China, including those traditional and emerging brands, for consumers to choose freely by their own preference in family gatherings or more significant occasions. What's more, Baijiu as a cultural carrier, actively promoted by some well-known cultural brands or media, is a strikingly typical way to share Chinese dietary culture with the world. Therefore, we should seize the opportunity to increase the overseas sales of famous Baijiu brands and expand the global market shares, so as to enhance the understanding of Baijiu and China in the western society.

Section A

The Selection of Famous Baijiu

Baijiu, noted for its *unique* aroma and taste, is one of the most local *features* in China. With various aroma, it could be obviously *identified* and carefully remarked by the ordinary people. To be precise, Baijiu has been *classified* into different aroma-based styles in more scientific ways to meet the need of consumers, produced by various processes in different regions of China. Today, there are three major categories of Baijiu, including sauce aroma, strong aroma, and light aroma. Apparently, the characteristics of these three aromas are more typical and representative than some minor categories, accounting for 60% to 70% of Baijiu in China *approximately*.

These three major types are represented and characterized by some famous brands; and these brands in turn benefit the development of those types of aroma in a long term. It is commonly believed that brand is a driving force to constantly enrich the knowledge of Baijiu. For this reason, knowing those famous brands of Baijiu is quite essential to the public. Different types of famous Baijiu will be selected from all brands in the market, which will *demonstrate* their respective characteristics.

1 Sauce Aroma

The first type is sauce-based aroma that is a highly fragrant distilled sorghum liquor of bold character and provides a flavor *resembling* Chinese fermented bean pastes and soy sauces. Actually, it was formerly called as "*Mou-aroma*" (茅香) because of the best-known Baijiu, Moutai. That is why Moutai has been considered as the founder of sauce aroma. Thus, when it comes to sauce aroma, Moutai is the most perfect example to demonstrate it.

Moutai is a world-famous brand of Baijiu with a unique "sauce fragrance", made in the town of Moutai near the the city of Renhuai in southwestern China's Guizhou province. It is one of the three famous distilled liquor in the world, together with *Scotch Whisky* and *French Cognac*. And Moutai also generally enjoys high reputation for its position, national liquor of China, because of its long history.

In terms of natural conditions in the town, it has unique advantages in Baijiu-brewing, its climate and vegetation in particular, lying solid foundation for brewing. With distinctive and *systematized* manufacturing techniques, Moutai is distilled from fermented sorghum that is mainly picked out of glutinous sorghum, commonly known as *Red Tassel Sorghum*. Within the string of processes, it is of significance to understand the "Distillation". It means that the fermenting mixture which composes Qu and sorghum is distilled seven times during one year, with each batch stored in a separate container. In fact, the aroma of the liquor need to change slightly under a certain seasonal conditions. Moutai now comes in different versions ranging from the standard 53% to 35% in *alcohol content*.

Moutai has a production history of over 200 years. During the Qing Dynasty (1644–1912), distillers from northern China introduced advanced techniques into local production processes to create a distinctive type of Baijiu, which was the origin of Moutai. Thereafter, Moutai was produced at several local distilleries. Moreover, during the Chinese Civil War, People's Liberation Army forces chose to camp at Moutai and then took some measures to manage the local liquor. Due to the Communist victory in the war, the government *consolidated* the local distilleries into one integrated company, and finally started Guizhou Moutai, which is a partial publicly traded and partial state-owned enterprise. It was praised as "National Liquor" in 1951, two years after the founding of People's Republic of China. Gradually, it has become a more popular option for people.

Moutai was known to the world after winning a gold medal at the 1915 *Panama-Pacific Exposition* in San Francisco, California. Meantime, it also claimed two gold medals separately at the Paris International Exposition in 1985 and 1986. By increasingly receiving wide exposure in China and abroad, Moutai has been a kind of "first choice" to make westerners get better understanding of China. It is one of the Baijiu brands usually served in China's official *state banquets* to heads of foreign states and distinguished guests visiting China, and it is even the only liquor presented as an official gift by Chinese embassies in foreign countries and regions. For example, during the state banquet for the U.S. presidential visit to China in 1972, Zhou Enlai used the liquor to *entertain* Richard Nixon. Premier Zhou told Nixon that Moutai had been famous since it won recognition in 1915, and that during the Long March, "Moutai was used by us to cure all kinds of diseases and wounds." Nixon replied, "Let me make a toast with this *panacea*." Honored as the "drink of diplomacy", Moutai maintains the most popular brand of Baijiu among Chinese political and business leaders for official receptions or other events.

Besides, Moutai currently sells over 200 tons of Moutai to over 100 countries and regions across the world. It is the world's most valuable liquor company, surpassing Diageo since April 2017. Thus, Guizhou Moutai is also working on global expansion plans.

2 Strong Aroma

Strong aroma is the second type to show the difference from other varieties of Baijiu. It refer to distilled liquor that is sweet, *unctuous* in texture, and *mellow*, with a gentle lasting fragrance contributed by the high levels of esters, primarily *ethyl acetate*, which give the spirit a strong taste of pineapple, banana and anise. Unlike sauce aroma, liquor of strong aroma has an obvious feature that it is distilled from sorghum, sometimes in combination with other grains, and continuously fermented in mud pits. In fact, strong aroma has also another name, formerly known as "Lu-aroma", because it was created by the Luzhou Laojiao Distillery in Luzhou. Thus, Luzhou Laojiao can completely reveal its characteristics. And there are other remarkable examples of this type of Baijiu is Wuliangye from Yibin, Jiannanchun from Mianzhu, and Yanghe from Suqian.

Luzhou Laojiao is one of the most popular liquor in China, with history extending over 400 years. Particularly, the history of its distillery can be dated back to the Qin and Han Dynasties, and it thrived in 1573 of the Ming Dynasty. Originated from a group of oldest liquor workshops in ancient times, Luzhou Laojiao obviously becomes one of the oldest Chinese liquor that is still in production. According to many well-recorded historical materials, Luzhou Laojiao could be often seen as the birthplace of Chinese Baijiu, now exclusively produced by Luzhou Laojiao Company Limited which is a large *state-owned enterprise* headquartered in Luzhou, Sichuan province, China. And the city of Luzhou has been won a great reputation, also known as the City of Baijiu.

As the typical representative of strong aroma of Baijiu, Luzhou Laojiao was firstly brewed with "thick fragrance" to be its foundation. It has the excellent traditional brewing technique that was listed as a National Intangible Cultural Heritage in 2006. Moreover, Luzhou Laojiao has a very famous symbol to be easily recognized from many liquor, that is, National Cellar 1573, which was originated from a group of brewing pits established in 1573. So far, Luzhou Laojiao has been stationary alcohol content ranging from 42% to 52%.

Benefited by the superior conditions, including climate and water source, Luzhou Laojiao is so renowned for the quality of its distillation along with its unique aroma and *mouthfeel*. Specifically, it is brewed in an environment with a unique clay composition that gives the liquor its famous aroma and palette. Luzhou, with a mild climate which contains an extreme temperature ranging from 40.3℃ to -1.1℃, annually abundant rainfall, and average annual wind speed, becomes the best place to breed regional unique crop quality and microbial groups. In other words, the native glutinous red sorghum better nourished by the distinctive climate can be carefully selected to be raw material of Luzhou Laojiao. However, apart from the proper climate, water source is the other part to make better liquor. Through a series of scientific analysis, the brewing water came from the Yangtze River has some noticeable characteristics, such as the water odorless, slightly sweet, was weakly acidic, appropriate degrees and sufficient trace elements. Exactly, these characteristics can effectively promote yeast breeding, favorable *glycosylation* and fermentation. Especially, the *enzymatic reaction* can be greatly increased in the production of daqu liquor.

With a strong aroma of fermented peaches (more than 50% alcohol), Luzhou Laojiao has been one of the best spirits at home and abroad and even one of China's oldest four famous liquor. For Luzhou Laojiao, brisk sales of its signature product, the National Cellar 1573, have been driving overall sales growth. And the company said total sales revenue would be likely to increase more than 25 percent year-on-year in 2018, mainly driven by growth in sales of high-end products. Therefore, Luzhou Laojiao pays more attention to its high quality being a more effective approach for sales, as its website states, "Our philosophy is to keep perfection of production and try our best to make consumers drink high quality of Baijiu."

Another remarkable representative of strong aroma type of Baijiu is Wuliangye, an aged distilled liquor produced in the city of Yibin in southern Sichuan province. Its brewing water primarily comes from the middle of the Min River. About its origin of history, it can be dated back to those significant periods of the Southern and Northern Dynasties (420A.D.-589A.D.). From the perspective of Chinese character, Wuliangye literally refers to five grains, while it exactly contains the five main ingredients in it, including sorghum, corn, glutinous rice, long-grain rice, and wheat. Selecting these five grains as raw materials, the formula was at least created in the Ming Dynasty and the name of Wuliangye was coined in 1905. The oldest cellars are reserved and still in use for more than 600 years. Gradually, Wuliangye is more nationalized and standardized, with the average alcohol content from 35% to 68%.

Wuliangye adopts *conventional* techniques in its brewing process and uses five well-selected grains into its fermentation. It has a unique and comprehensive style of long-lasting fragrance, mellow and rich taste, clean and pure flavor entering the throat, and balanced flavors in harmony, for it owns a number of materials. Then, Wuliangye, with peculiar ecological conditions, wins great reputation for these characteristics and receives more attention from the domestic and overseas market. Besides, it stresses that keeping the tradition is the foundation to produce high-quality liquor, *renowned* as currently the outstanding treasure of various liquor.

On account of these advantages, Wuliangye has claimed the gold medals of "National Liquor" for many times. In 1991, it was awarded as "Ten Famous Trademarks" in China. And then, after winning a gold medal for 80 years in the 1915 Panama-Pacific Exposition, Wuliangye once again received the only gold medal at the International Exposition in 1995 and in 2002 respectively. It was crowned as "The King of Chinese Liquor Industry" due to the medal in 1995, and consequently it established the splendor and achievement. So far, Wuliangye has gained countless honors and awards in a wide variety of world's exhibitions or competitions, and continued to write down its legendary history with unique production.

As a high-quality brand, Wuliangye is increasingly becoming global and its growth has become a "standout" story. According to an online report released by the London-based market research company in China Daily, it could be obviously found that Wuliangye was the fastest-growing brand of 2018 and the spirit industry champion, rising 184 places to 100th with a brand value of $14.6 billion, up 161 percent year-on-year. Besides, Wuliangye Yibin Co. Ltd. said it sold 6,000 metric tons of its liquor across product categories in the first quarter of 2018, higher than that in the past few years on a quarterly basis. Recognized worldwide as quality brands, Wuliangye Yibin Co. Ltd. aims at constructing a more newly innovative pattern of industry, which puts liquor-centered business into focus and achieves multiple development, such as manufacturing with modern machine, material science, modern packaging and logistics.

3 Light Aroma

As the third type of aroma, light aroma has a number of outstanding attributes: delicate, dry, and mild, with a delectable mellow and clean texture in the mouth. More precisely, it can give mellow sweetness and refreshing *aftertaste*. These flavors of the distilled liquor are contributed primarily by a balance between ethyl acetate and *ethyl lactate*, and give the spirit a taste of dried fruit with floral notes. And then, it is made from sorghum fermented in a stone vessel with qu made from wheat bran or a combination of barley and peas. There are the two typical models to perfectly present this liquor, Erguotou from Beijing and Fenjiu from Shanxi. The former is known as Kaoliang in Taiwan. However, the latter, in fact, is the best example because the light aroma of Baijiu made in the Xinghuacun Fenjiu Distillery was called "Fen-aroma".

Fenjiu is distilled sorghum-based Baijiu originally produced in Xinghuacun town, Fenyang, Shanxi province. Fenjiu has a long history of about 4,000 years to be the original Chinese sorghum Baijiu, so that it had three glorious periods in the Chinese history. It was first dated back to the Northern and Southern Dynasties (550A.D.), recorded into the *Twenty-Four Histories* for gaining the popularity of the Emperor of the Northern Qi Dynasty. And then, Xinghuacun was featured in the poem "Qing Ming" written by the great poet Du Mu (803A.D.-852A.D.) in the late Tang Dynasty, which is now a part of mandatory reading for primary school. Exactly, due to the famous poem, Fenjiu gained great fame, also called as "Xinghuacun Liquor". Finally, in 1915, Fenjiu won a gold medal at the 1915 Panama-Pacific International Exposition, representing a milestone in the development of Chinese Baijiu and a leader of Chinese Brewing Industry. Meantime, the long history of Fenjiu has the same origin with Chinese civilization, the culture of Yellow River, and the culture of Shanxi Merchants, so that Fenjiu can develop its own aroma characteristics among all liquor and one of the most competition-based advantages, generally *acknowledged* by the community of Baijiu. As the national treasure, Fenjiu is an integration of the wisdom from Chinese people in ancient times.

Among a wide variety of Baijiu, Shanxi Fenjiu is the originator of the typical light aroma, and even becomes a best model of the light-aroma style based on the national standard of Baijiu industry. It is also the most popular local liquor that is generally sweeter than other liquor made in northern China. Being a *prominent* representative of light aroma style, Fenjiu has a deep tradition and superior techniques to make people enjoy long-lasting and sweet taste, and it can give rise to a long-remaining flavor after drinking. More precisely, by selecting high quality local sorghum, barley and pea as raw material from the middle of Shanxi province, most skilled masters in brewing usually create a distinctive agent of *glycation fermentation* made from barley and pea and adopt a specialized technique of "*double separating distilling raw and fermented material*". Correspondingly, it can be noticeable that the unique and exact characteristics of Fenjiu, with alcohol content ranging from 38% to 65%, are pure and mild flavor, mellow sweetness and refreshing aftertaste.

Besides, those talented craftsmen with highly practical skills give their peculiar insight into the whole process, from qu-making to fermentation, and then to distillation. For hundreds of years, these skills have been passed down from generation to generation by oral communication of *master-apprentice relationship*, and have been increasingly innovated and developed themselves.

Another factor is the water source that plays an essential role in brewing Fenjiu. Water is the content of liquor, while qu is the main structure. Naturally, there is an old saying that "where there has a wonderful spring, there must have a famous liquor." From the modern science, it can be revealed that the secret of Fenjiu comes from the abundant and high-quality groundwater resources, containing various chemical elements which are *beneficial* to people's health and brewing process. Thus, Mr. Fu Shan, who was a famous poet, calligrapher, and physician of traditional Chinese medicine in the Qing Dynasty, wrote an *inscription* to think highly of those health benefits from Fenjiu. As a result, Fenjiu is well-known for its excellent techniques of both traditional crafts and modern science and purified water source of high quality.

Currently, Fenjiu has been a cultural icon of not only Shanxi Baijiu industry but also China. Fenjiu became famous 1,500 years ago, and it gradually forms its own liquor-producing area supported by government. With the development of Shanxi's economy and tourism, Fenjiu establishes a cultural scenic area which features an ancient-looking *cluster* of buildings that covers a floor space of 1.58 million square meters, the size of ten Palace Museums. Over 100 imposing black-tiled buildings stand close to each other and were fashioned based on elements from the Dynasties of Tang, Song, Ming and Qing. In the museum, visitors can sample beverages and see the process of liquor making from brewing to storage, filling and packaging. The Fenjiu Cultural scenic area, with its brewing techniques recognized as a national intangible cultural heritage, has supported the development of the local liquor industry and industrial tourism.

New Words and Expressions

1 unique [juːˈniːk] *adj.* 独特的，稀罕的
2 feature [ˈfiːtʃə] *n.* 特色，特征
3 identify [aɪˈdentɪˌfaɪ] *vt.* 确定；鉴定
4 classify [ˈklæsɪfaɪ] *vt.* 分类
5 approximately [əˈprɒksɪmətlɪ] *adv.* 大约，近似地
6 demonstrate [ˈdemənstreɪt] *vt.* 展示
7 resemble [rɪˈzembl] *vt.* 类似，像
8 systematize [ˈsɪstɪmətaɪz] *vt.* 使系统化；使组织化；将……分类
9 Red Tassel Sorghum 红缨子高粱
10 alcohol content 酒精含量
11 consolidate [kənˈsɒlɪdeɪt] *vt.* 联合
12 Panama-Pacific Exposition 巴拿马太平洋博览会
13 state banquet *n.* 国宴
14 entertain [ˌentəˈteɪn] *vt.* 招待
15 panacea [ˌpænəˈsiə] *n.* 灵丹妙药；万能药
16 unctuous [ˈʌŋktjʊəs] *adj.* 油质的
17 mellow [ˈmeləʊ] *adj.* 芳醇的

18 ethyl acetate [有化] 乙酸乙酯
19 state-owned enterprise 国有企业；国企
20 mouthfeel [ˈmaʊθfiːl] n. 口感
21 glycosylation [ˌɡlaɪkəsɪˈleɪʃən] n. 糖基化
22 enzymatic reaction 酶反应；酶促反应
23 conventional [kənˈvenʃənəl] adj. 符合习俗的，传统的
24 renowned [rɪˈnaʊnd] adj. 著名的
25 ethyl lactate [有化] 乳酸乙酯
26 acknowledge [əkˈnɒlɪdʒ] vt. 承认
27 prominent [ˈprɒmɪnənt] adj. 突出的，显著的
28 glycation fermentation 糖化发酵
29 double separating distilling raw and fermented material [专] 清蒸二次清
30 master-apprentice relationship 师徒关系
31 beneficial [ˌbenɪˈfɪʃəl] adj. 有益的，有利的
32 inscription [ɪnˈskrɪpʃən] n. 题词
33 cluster [ˈklʌstə] n. 群

Exercises

I. Answer the following questions.

1 What are the major aromas of Baijiu in China?
2 What does the sauce-aroma style Baijiu feature?
3 What are the natural conditions that determine the distinctive style of Moutai?
4 Why is Moutai considered as a cultural symbol of China?
5 What is the perfect example of Luzhou Laojiao?
6 From the literal meaning, what does Wuliangye refer to?
7 What are the characteristics of Wuliangye?
8 What are the three glorious periods of Fenjiu in the Chinese history?

II. Discuss the following questions.

1. Give a brief introduction of the superior conditions for Luzhou Laojiao.
2. Tell some historical stories about Fenjiu.

III. Match the items in column A with those in column B.

A	B
1. 国宴	____a. mouthfeel
2. 酒精含量	____b. National Cellar 1573
3. 师徒关系	____c. Panama-Pacific Exposition
4. 浓香型	____d. aftertaste
5. 回味	____e. master-apprentice relationship
6. 国酒	____f. state-owned enterprise
7. 国窖1573	____g. alcohol content
8. 巴拿马太平洋博览会	____h. National Liquor
9. 口感	____i. Strong aroma
10. 国有企业	____j. state banquet

IV. Fill in the following blanks with the words given below. Change the form when necessary.

| classify | feature | entertain | renowned | superior |
| resemble | consolidate | beneficial | conventional | unique |

1. He very much _____ a friend of mine.
2. This magazine will be running a special _____ on education next week.
3. The books in the library are _____ by subject.
4. He made a few _____ remarks about the weather.
5. The match will show who is the _____ player.
6. A museum should aim to _____ as well as educate.
7. The region is _____ for its outstanding natural beauty.
8. The city has an atmosphere which is quite _____.

9 The arrangement was mutually _____.
10 With his new play he has _____ his position as the country's leading dramatist.

V. Translate the following passage into English.

As the superior core of Chinese Baijiu, there have two well-known national liquor, Moutai and Dongjiu, in Guizhou province. Of the two brands, entitled in 1942 has owned a popularity from the market of Baijiu. In the industry of distilled liquor, Dongjiu has gradually developed its own features from raw materials to processing techniques, particularly the prefect combination of Daqu and Xiaoqu. Inheriting the deep foundation of ancient Chinese culture, Dongjiu is considered as a national treasure and truly living fossil in the tradition of Chinese medicine from thousands of years, with most-distinguishing sign of China.

In 1980s, Dongjiu was comparable in the competition of high-quality Baijiu along with Moutai, Wuliangye, and Luzhou Laojiao. Thus, since 1963, it has won many awards or gold medals in the annual conferences whose aim is to taste and evaluate the quality of Baijiu in China, and even got world-class honors overseas. And Dongjiu is permanently classified as "Secret of State" by the national authority, because of its unique feature and originally scientific formula. In 2008, it was officially recognized as the local standard and the most representative of Medicine-aroma Baijiu.

VI. Write a passage according to the given outline with no less than 150 words.

A Brief Introduction on Wuliangye

1）五粮液的起源；
2）五粮液的特点；
3）五粮液的地位与影响。

Section B

The Minor Styles of Baijiu

[A] As is well-known, there are *numerous* styles of Baijiu in China, in particular three major aroma-based categories. According to some recent studies *conducted* by economist and researchers, it could also be demonstrated that these three flavor types are very typical and representative and account for a large-scale share of Baijiu market. The production techniques for these three types of Baijiu, including sauce-aroma, strong-aroma, and light-aroma, are standardized and *stereotyped*. In addition to these three types, some styles of Baijiu, such as Sanhuajiu, Xifengjiu, and Baiyunbian, with a specific flavor and aroma characteristics are also produced with different techniques. These Baijiu do not belong to the first three categories, but have their own flavors, which are described as follows.

1 Rice Aroma

[B] Regarded as one of the four traditional liquor in China, rice-aroma Baijiu is characterized by soft sweet flavor and clean aftertaste and thus also called as "honey aroma". This type of Baijiu contains more content of higher alcohol, dominated by ethyl lactate which is higher than ethyl acetate. Selecting high-quality rice as raw material and making xiaoqu starter for fermentation, rice-aroma Baijiu adopts semisolid-state fermentation as its main production technique without any *edible alcohol* or other *substances* not fermented by Baijiu. The liquor distilled from rice has a typical example, Sanhuajiu, "Three Flowers Liquor", with *allegedly* over a thousand-year history made in Guilin, Guangxi province. It is famous for the fragrant herbal addition, and the use of spring water from Mount Xiang in the region. Its alcohol content is 55%–57%.

[C] According to the latest standard, rice-flavor Baijiu is divided into two sub-categories, primarily including traditional type and modern type. The traditional type is represented by Sanhuajiu from Guilin, while the modern one has a specific example, Bingyu Zhuangyuan made from the original rice-fermented Baijiu. In fact, the *division* can be seen as a demand of economic growth and a product of scientific innovation. The new one not only keeps the tradition of rice-aroma Baijiu, but also seek to meet the demand of modern *consumption* for various health effects, such as *nutrition*, *food therapy*, high quality, etc. However, it can be noticeable that rice liquor is entirely different from rice-aroma liquor. The former is mainly *home-brewed* from rice with a certain sediment at normal temperature, while the latter operates in the factory or workshop and adopts certain processing technique to *eliminate* all *precipitating* substances. The difference can be observed just with naked eyes.

2 Phoenix Aroma

[D] Phoenix aroma refers to a type of distilled liquor fermented in mud pits and aged in *rattan* containers. All liquor of this aroma have a fruity taste similar to strong-aroma Baijiu, as well as an *earthier* quality and an expanding finish. Generally speaking, the liquor is chiefly made from sorghum, barley and pea as its raw materials. It adopts medium-temperature Daqu or Fuqu together with yeast to make *secchariferous and fermentative agent*. And then it adopts a mixed distilled order way to put into compounds in the mud pit whose age is not more than a year. Its main ingredients of flavor are from a balance among ethyl acetate, *ethyl hexanoate*, and *isoamyl* alcohol, showing colorless and *transparent* appearance, *ethanol* aroma, mellow and clean mouthfeel, clean purity, and balanced flavors. An example of phoenix-aroma style Baijiu is Xifengjiu from Fengxiang County in Shaanxi, which is characterized by sweet entrance and elegant aftertaste.

[E] Xifengjiu boasts a long history becoming one of the oldest liquor in China, first brewed over 3,000 years ago in the Shang Dynasty and flourished in the Tang Dynasty. According to Chinese *folklore*, Fengxiang was the home to phoenix, thus the name Xifengjiu means Baijiu produced in west of phoenix home. It was awarded as "National Liquor" four times in the official evaluation among Baijiu industry. In recent years, Xifeng has put great emphasis on innovation and product development, making great breakthrough in production techniques. On basis of its traditional techniques and flavor, Xifeng strives to develop various series of products with unique taste and different levels of alcohol, *catering* to different needs of customers in different areas. Meantime, through detailed analysis of its micro *compositions*, it could be demonstrated that Xifengjiu belongs to neither mild-aroma style nor strong-aroma style, having a unique flavor itself.

3 Mixed Aroma

[F] Mixed-aroma Baijiu, also known as *Miscellaneous*-flavor Baijiu, refers to a type of distilled liquor that is a blend of two or more varieties of Baijiu, influenced by natural conditions, raw materials and brewing techniques. Generally, this type of Baijiu means that it can randomly combine at least two kinds of liquors, such as sauce-aroma style, strong-aroma style, light-aroma style, to organically form a completely new kind. Therefore, it has *sensory* characteristics ranging between those of sauce-flavor and strong-flavor Baijiu. The representative aroma compounds in miscellaneous-flavor Baijiu are *heptanoic acid*, *ethyl heptanoate*, *isoamyl acetate*, *2-octanone*, *isobutyric acid*, and *butyric acid*. To satisfy the needs of scientific diet and mouthfeel to modern people, it aims at improving the unbalanced aftertaste of sauce-aroma Baijiu and removing the very heavy mouthfeel of strong-aroma Baijiu. Then, it has recently enjoyed great popularity of consumers along with the two other kinds—sauce-flavor Baijiu and strong-flavor Baijiu.

[G] As the best example of this type, Baiyunbian literally means "Next to the White Cloud Liquor", produced in Songzi, Hubei province in Central China. Established in 1952, its company has become a well-known brand of Chinese liquor and even a model in efforts to win the market by quality over the past 60 years. And it adopts Chinese traditional production techniques, so that it has liquor storage of 70 thousand tons at its warehouse. The main characteristic of Baiyunbian is a combined style, which is sauce fragrant flavor containing heavy or thick fragrant flavor. Actually, as a representative of mix-flavored Chinese liquor, with its aroma close to that of Moutai. In addition, according to a recent report in China Daily in 2016, the *revenue* from sales of Baiyunbian reached 4.35 billion yuan ($653 million), with a tax quota of 790 million yuan. Thus, it has gained high praise and *preference* from the masses across the nation, rather than only in Hubei province.

New Words and Expressions

1. numerous [ˈnjuːmərəs] *adj.* 许多的，很多的
2. conduct [ˈkɒndʌkt] *vt.* 实施；执行
3. stereotype [ˈsterɪəˌtaɪp] *vt.* 使一成不变
4. edible alcohol *n.* 食用酒精
5. substance [ˈsʌbstəns] *n.* 物质
6. allegedly [əˈledʒɪdlɪ] *adv.* 据说，据称
7. division [dɪˈvɪʒ(ə)n] *n.* [数] 除法；部门；分配；分割
8. consumption [kənˈsʌmpʃən] *n.* 消费

9 nutrition [njuːˈtrɪʃən] n. 营养，营养学
10 food therapy 食疗
11 home-brewed [ˈhəʊmˈbruːd] adj. 自酿的
12 eliminate [ɪˈlɪmɪˌneɪt] vt. 消除
13 precipitate [prɪˈsɪpɪˌteɪt] n. [化学] 沉淀物
14 Phoenix aroma 凤香型
15 rattan [ræˈtæn] n. 藤
16 earthy [ˈɜːθɪ] adj. 朴实的
17 sacchariferous and fermentative agent 糖化发酵剂
18 ethyl hexanoate 已酸乙酯（一种增香剂）
19 isoamyl or isoamyl acetate [有化] 乙酸异戊酯
20 transparent [trænˈspærənt] adj. 透明的
21 ethanol [ˈeθəˌnɒl] n. [有化] 乙醇，[有化] 酒精
22 folklore [ˈfəʊkˌlɔː] n. 民间传说
23 composition [ˌkɒmpəˈzɪʃn] n. 成分
24 miscellaneous [ˌmɪsəˈleɪnɪəs] adj. 混杂的
25 sensory [ˈsensərɪ] adj. 感觉的
26 heptanoic acid [有化] 庚酸
27 ethyl heptanoate 庚酸乙酯
28 2-octanone [ɒktəˈnɒn] n. 甲基己基甲酮
29 isobutyric acid n. [有化] 异丁酸
30 butyric acid n. [有化] 丁酸；酪酸
31 revenue [ˈrevəˌnjuː] n. 收益
32 preference [ˈprefərəns] n. 偏爱，倾向

Exercises

I. Reading comprehension.

The following ten statements are derived from the passage. Each statement contains information given in one of the paragraphs. Identify the paragraph from which the information is derived. Each paragraph is marked with a letter.

1 Semisolid-state fermentation is the main production technique of rice-aroma Baijiu adding no edible alcohol or other substances not fermented by Baijiu.
2 Miscellaneous-flavor Baijiu refers to a type of distilled liquor influenced by natural conditions, raw materials and brewing techniques, randomly combining at least two kinds of aromas to organically form a completely new kind.
3 From the literal meaning, the brand Baiyunbian can be actually explained "Next to the White Cloud Liquor".
4 Mixed-aroma Baijiu has a more specific goal to improve the unbalanced aftertaste of sauce-aroma Baijiu and remove the very heavy mouthfeel of strong-aroma Baijiu, so that it enjoys great popularity among consumers along with the two other kinds.
5 To satisfy the demand of economic growth and transform scientific innovation, rice-flavor Baijiu is divided into traditional type and modern type.
6 Sanhuajiu, Xifengjiu, and Baiyunbian are other styles of Baijiu with their own specific flavor and aroma characteristics rather than three major aroma-based categories.
7 Phoenix-aroma Baijiu takes medium-temperature Daqu or Fuqu as its starter, and sacchariferous and fermentative agent was made from added substance.
8 Originated in the Shang Dynasty and flourished in the Tang Dynasty, Xifengjiu boasts a long history to become one of the oldest liquor in China
9 Sanhuajiu is a typical sign to demonstrate the liquor distilled from rice, known for the fragrant herbal addition, and the use of spring water from Mount Xiang in the region.
10 Xifengjiu, made from Fengxiang County in Shaanxi province, is an outstanding representative of Phoenix-aroma Baijiu.

II. Translate the following passage into English.

剑南春产于四川省绵竹市，是国内浓香型白酒的典型代表之一。绵竹素有"酒乡"之称，且产酒而得名。同时它也是川酒发源地之一。因其独特而优越的自然条件，为剑南春提供了较好的酿酒环境。除此之外，绵竹酿酒历史悠久，源远流长，酒文化特别丰富。从绵竹的史料和文物可以看出，延续2400年的剑南春酒不仅是四川酒史的重要组成部分，也是我国珍贵的文化遗产。

课后练习答案及课文译文

Unit 1

Section A

I.

1. Liquor-making enjoys a very long history in China. China is one of the countries which first made liquor in the world.
2. In the history of Chinese civilization, liquor-making and Chinese culture were synchronized.
3. Chinese people began making liquor about 40,000-50,000 years ago in the Paleolithic Period.
4. They imitated the biological winemaking process of nature.
5. The purposeful Liquor-making began in China after mankind entered the Neolithic Age and with the emergence of agriculture.
6. The oldest wine in the world is the wine unearthed in Samari, Iran. The oldest liquor in China is the imperial liquor of the Han Dynasty unearthed in Xi'an.
7. Yes, there are numerous lines about liquor and liquor-making in historic books and literary works. For example, there is such a line in *The Book of Songs* as "I am drunk with your liquor and enlightened by your virtue", in *The Book of Rites* as "Liquor can prolong one's life", and "Liquor is used to show etiquette and you cannot drink too much" in *The Book of Zuo Zhuan*.
8. The earliest wines in China were made from fruits and rice.
9. Chinese people invented the distillation method in the Song Dynasty.
10. They are Moutai, Wuliangye, Jiannanchun, Luzhou Laojiao Tequ, Xifengjiu, Fenjiu, Gujing Gongjiu, Dongjiu, Yanghe Daqu and Langjiu.

III.

1 g   2 h   3 j   4 i   5 d   6 f   7 e   8 a   9 c   10 b

IV.

1 enlightened;   2 ubiquitous;   3 prevails;   4 permeates;
5 germinate;   6 vicissitudes;   7 prolonged;   8 unscrupulous;
9 brew;   10 aromatic

V.

Yellow wine, also called rice wine, is a Chinese specialty. It is one of the three brewed liquor of the world, the other two are wine and beer. The technologies involved in brewing yellow wine are the typical representative and role model of the eastern brewing world. The Shaoxing Yellow Wine made from rice is the oldest and most representative product. Different from the spirits, people get yellow wine not through distillation but through brewing and its content of alcohol is less than 20%. Different types of yellow wine come in different colors. Some are beige, some are brownish yellow and some are reddish brown. Jimo Laojiu of Shandong Province is the typical representative of millet yellow wine of north China. Chengangjiu of Longyan, Fujian Province and Fujian Laojiu are the typical representatives of red-yeast yellow wine.

译文

中国酿酒简史

中国是酒的故乡，也是酒文化的发源地，是世界上最早酿酒的国家之一。酒的酿造在中国已有相当悠久的历史。在中国数千年的文明发展史中，酒与文化的发展基本上是同步进行的。在中国，酒的起源可以追溯到史前时期。最初的酒是含糖物质在酵母菌的作用下自然形成的有机物。自然界中生长着大量的含糖野果，空气里、尘埃中和果皮上都附着有酵母菌。在适当的水分和温度等条件下，酵母菌就有可能使果汁变成酒浆，自然形成酒。中国人酿酒的历史约始于距今4万—5万年前的旧石器时代。当时人们有了足以维持基本生活的食物，从而有条件去模仿大自然生物本能的酿酒过程。人类最早的酿酒活动只是机械地简单重复大自然的自酿过程。真正称得上有目的的人工酿酒生产活动，是在人类进入新石器时代、出现了农业之后开始的。这时，人类有了比较充裕的粮食，尔后又有了制作精细的陶制器皿，这才使得酿酒生产成为可能。公元前5000年—前3000年，中国仰韶文化时期已出现耕作农具，即出现了农业，这为谷物酿酒提供了可能。《中国史稿》认为，仰韶文化时期是谷物酿酒的"萌芽"期。当时是用糵（发芽的谷粒）造酒。公元前2500年—前2000年的中国龙山文化遗址出土的陶器中，有不少尊、盉、高脚杯、小壶等酒器，反映出酿酒在当时已进入盛行期。

酒，是人类各民族民众在长期的历史发展过程中创造的一大饮料。世界上最古老的酒是伊朗撒玛利出土的葡萄酒，距今三千多年，仍芳醇迷人；中国最古老的酒是西安出土的汉代御酒，据专家考证系粮食酒，至今仍香醇可饮。中国甲骨文中早就出现了"酒"字和与酒有关的醴、尊、酉等字。从中可以看出酒的历史悠久。至于文史中的记载更是不胜枚举，如《诗经》中有"既醉以酒，既饱以德"的诗句。关于古代酒俗的记载更多，如"酒者可以养老也"（《礼记》）、"酒以成礼，不继以淫"（《左传》）等。这说明酒存在着多种用途。

中国制酒源远流长，品种繁多，名酒荟萃，享誉中外。最早的酒是果酒和米酒。自夏之后，经商周、历秦汉，以至于唐宋，皆是以果实、粮食蒸煮，加曲发酵，压榨而后才出酒的。随着人类的进一步发展，酿酒工艺也得到了进一步改进，在宋代，中国人发明了蒸馏法，从此，白酒成为中国人饮用的主要酒类。酿酒技术发展到今天，中国出现了品种繁多的酒类产品，如白酒、啤酒、葡萄酒、黄酒、米酒和药酒等。中国的十大白酒名酒包括：茅台、五粮液、剑南春、泸州老窖特曲、西凤酒、汾酒、古井贡酒、董酒、洋河大曲、郎酒。

数千年来，富有创造力的中国人在历史的变迁中，酿造出了许多更具有地方特色、更能反映当地风土人情的各类名酒，外加不同地域和不同民族的酒礼酒俗，无不显示中国是一个博大的名酒古国。酒渗透于整个中华五千年的文明史中，无论从文学或艺术创作、文化娱乐到饮食烹饪，还是在养生保健等各方面，酒在中国人生活中都占有重要的位置。

Section B

I.

1 D   2 C   3 C   4 D   5 C   6 B   7 D   8 B

II.

There is an interesting legend about wine in the West. It is said that an ancient Persian king sealed the surplus grape in a bottle and wrote the word "TOXIC" on the bottle for fear that his storage would be stolen. The king was too busy and soon he forgot his stored grape completely. One day an unloved concubine happened to see the bottle with the word "TOXIC" on it. She decided to end her life soon. Opening the bottle, she found the oddly colored liquid in the bottle seemed to be toxicant, so she took several swigs and lay there waiting to die. After a while, she did not feel unbearably painful but comfortable and drunk. She reported this odd thing to the king and the king felt greatly surprised. The king took some too and he felt the same as the concubine did. This is the origin of wine.

译文

西方酿酒起源

　　酿酒在西方没有明确的历史记载。多数史学家认为，西方酿酒是从酿造葡萄酒开始的，葡萄酒的酿造首先出现在波斯，后来传到了当时希腊的克里特岛，然后经过战争和通商才传到了法国等地。葡萄酒的酿造起源于公元前6000年的古代波斯，即现今的伊朗。

　　大约在公元前3000年，埃及人就已经开始酿造葡萄酒了。在埃及的古墓中所发现的大量的珍贵文物（特别是浮雕）清楚地描绘了当时古埃及人栽培、采收葡萄和酿造葡萄酒的情景。对于葡萄的最早栽培，大约是在7000年前始于南高加索、中亚细亚、叙利亚、伊拉克等地区。后来随着战争、移民传到其他地区。初至埃及，后到希腊。但是，真正可寻的资料，还是从埃及古墓中发现的大量遗迹、遗物。在尼罗河河谷地带，从发掘的墓葬群中，考古学家发现一种底部小圆，肚粗圆，上部颈口大的土罐陪葬物品，经考证，这是古埃及人用来装葡萄酒或油的土陶罐；特别是浮雕中，清楚地描绘了古埃及人栽培、采收葡萄、酿制葡萄酒步骤和饮用葡萄酒的情景，至今已有5000多年的历史。此外，埃及古王国时代所出品的酒壶上，也刻有伊尔普（埃及语，即葡萄酒的意思）一词。西方学者认为，这才是人类种植葡萄与酿造葡萄酒的开始。以葡萄酒为主题的著名作家休·约翰逊曾描写道："古埃及有十分出色的品酒专家，他们就像二十世纪的雪利酒产销商或波尔多酒经纪人一样，可以自信并专业地鉴定酒的品质。"

葡萄酒基本与人类文明共同起源，世界上古老的神话传说中都流传着关于葡萄酒的故事，葡萄酒的历史中也有人类历史发展的踪迹。

从用葡萄酿酒开始，人们不断尝试着用各种食材来酿酒，世界各地酒的种类有数万种，酿酒所用的原材料和酒精含量也有很大的差异，人们为了便于了解和记忆，于是就用不同的方法将它们予以分类。若以生产原料进行分类，大致可以分为谷物酒、香料草药酒、水果酒、奶蛋酒、植物浆液酒、蜂蜜酒和混合酒七大类；若以饮用时间来分类又有餐前酒（或称开胃酒）、佐餐酒、餐后酒和特饮酒。若以酒精含量的不同来分类可分低度酒、中度酒和高度酒。但若以酒的生产工艺来加以划分，可以分三类：1．发酵酒类。包括葡萄酒、啤酒、米酒和苹果酒等。2．蒸馏酒类。包括威士忌酒、白兰地酒、伏特加酒、兰姆酒、特其拉酒和中国白酒。3．精炼和综合再制酒。包括金酒、利口酒、味美思酒、苦味酒、竹叶青酒、人参酒等。在上诉各种分类中，较为普及和广泛运用的酒类划分方式是以酒的生产工艺来划分。

随着人类历史变迁，出现了很多世界名酒。例如象征着财富与地位的人头马路易十三（法国），被誉为兴奋和骚动之源的百家得朗姆酒（西印度群岛），燃烧着的生命之水伏特加（俄国），皇家饮品芝华士（苏格兰），白兰地神话轩尼诗（法国），流淌的黄金马爹利（法国），恬淡的乡村情怀杰克丹尼（美国）和怒然绽放的花火特其拉（墨西哥）。从人们给这些酒冠以的各具特色的称谓中，我们不难看出酒在西方人的日常生活、人际交往中都占据着重要地位。

Unit 2

Section A

I.

1 Apes put fruits into the hole and brewed them into wine unintentionally.
2 Because Da Yu realized that he would get addicted to it and it disturbed his work.
3 Because the trade with other countries had been largely increased.
4 There are lots of functions of Chinese Baijiu, such as making friends through enjoying Baijiu together, and sending it to others as gifts.
5 They all love to drink Baijiu.
6 Du Kang.
7 In that stage, Baijiu was treated as a magic drink and there was no detailed method to make Baijiu.
8 In this stage, the distiller in western regions was introduced to China, which contributed to the invention of Chinese Baijiu.

III.

a-7   b-9   c-1   d-3   e-10   f-8   g-5   h-2   i-4   j-6

IV.

1 take charge of	2 are lacking in	3 integral	4 be traced back
5 exaggerated	6 banquet	7 indispensable	8 discard
9 verified	10 beard much significance in		

V.

尧帝为上古五帝之一，传说他是真龙化身，对灵气特别敏感。他受滴水潭灵气所吸引，带领大家在附近安居，并发展农业，使得老百姓过上了安定的生活。为感谢上苍，尧帝精选出最好的粮食，并用滴水潭的水浸泡，去除杂质，取其精华，合酿祈福之水。此水清澈纯净，清香悠长，便是最早的酒。

译文

白酒的起源

中国白酒的历史悠长久远,甚至可以上溯到上古时期。《史记·殷本纪》就有关于纣王"以酒为池,悬肉为林","为长夜之饮"的记载。白酒一直影响着中国社会的发展,无论是在重大的宴会、社交礼仪还是在日常生活中都可以看到白酒的身影,因而白酒是中国历史文化中不可缺少的一部分。对于酒的起源时间并没有具体的记载时间,只是酒史的记载长于国史,于是民间流传着这样的说法"先有酒,后有国"。对于酒的起源更是有多种说法,一般情况下,大致分为三类:

1. 上天造酒说

上天造酒的说法主要源于古代民间的传说、文献记载和文人诗歌三种途径。《酒谱》中的说"酒,酒星之作也",中国人的祖先有认为"酒,就是酒神仙主管的"。文人的诗歌中再次证明这一点:素有"诗仙"之称的李白,在《月下独酌其二》一诗中有"天若不爱酒,酒星不在天"的诗句;在《与曹操论禁酒书》中有"天垂酒星之耀,地列酒泉之郡"之说;意思是自古以来,我国祖先就有酒是天上"酒星"所造的说法。然而,酒自"上天造"之说,既无立论之理,又无科学论据,此乃附会之说,文学渲染夸张而已。

2. 猿猴造酒说

根据一些远古化石的记载,猿猴不仅嗜酒而且还造酒,这在我国的许多典籍中都有记载。清代文人李调元在他的著作中记叙到"琼州(海南岛)多猿……尝于石岩深处得猿酒,盖猿以稻米杂百花所造,味最辣,然极难得。"明代文人李日华在他的著作中,也有类似的记载,"黄山多猿猱,春夏采杂花果于石洼中,酝酿成酒,香气溢发"这些不同时代不同人的记载,足以可以证明这样的事实,即在猿猴的聚居处,多有类似"酒"的东西发现。即猿猴将果实丢弃或储藏在石头缝或者树洞里,其糖分自然发酵成酒浆。然而,这些说法缺乏翔实的考证,故不具备严格的文化意义。

3. 仪狄、杜康造酒说

相传夏禹时期的一位绝色女子仪狄发明了酿酒,公元前二世纪史书《吕氏春秋》云:"仪狄作酒"。相传她在做饭时闻到谷物的芳香,便将所其发酵所成的汁液呈与大禹品尝。大禹喝后觉得甘甜无必,并连饮数杯,逐渐上瘾,导致政事耽搁。

后来大禹意识到喝酒误事，便刻意戒酒，并且远离仪狄，还预言将来必有人因酒亡国。这就和汉代刘向编辑的《战国策》吻合。史籍中多处提到仪狄乃制酒之始祖。另一种说法是杜康也是制酒之始祖。关于杜康的神秘身份，都有很多种说法。有一种说法是杜康是少康，乃是夏王朝的第六位君王；另一种说杜康是周代人。根据历史的记载，应该是夏朝的说法比较靠谱。传说杜康将未吃完的剩饭剩菜放置在桑园的树洞中，经发酵后其散发出芳香气味。这就是传说中的酿造方法，因此人们就把杜康当作酿酒的祖先，"杜康"也成为酒的代名词。

总之，在几千年漫长的历史过程中，中国传统酒呈段落性发展。公元前4000-2000年，即由新石器时代的仰韶文化早期到夏朝初年，为头一个段落。这个时期是原始社会的晚期，先民们无不把酒看作是一种含有极大魔力的饮料。这个段落，经历了漫长的两千年，是我国传统酒的启蒙期。用发酵的谷物来酿造水酒是当时酿酒的主要形式。从公元前2000年的夏王朝到公元前200年的秦王朝，历时一千八百年，这一段落为我国传统酒的成长期。就在这个时期，酿酒业得到很大发展，并且受到重视，官府设置了专门酿酒的机构，酒由官府控制。第三段落由公元前200年的秦王朝到公元1000年的北宋，历时一千二百年，是我国传统酒的成熟期。饮酒不但盛行于上层社会，而且普及到民间的普通人家。这一段落的汉唐盛世及欧、亚、非陆上贸易的兴起，使中西酒文化得以互相渗透，为中国白酒的发明及发展进一步奠定了基础。第四段落是由公元1000年的北宋到公元1840年的晚清时期，历时八百四十年，是我国传统酒的提高期，其间由于西域的蒸馏器传入我国，从而导致了举世闻名的中国白酒的发明。而中国白酒则从此欣欣深入生活，成为人们普遍接受的饮料佳品，不断有的创新和发展适应人们的需求。

Section B

I.

1 B 2 A 3 A 4 B 5 A 6 C 7 A 8 D

II.

Chinese people began to make liquor with grains 7,000 years ago. Generally speaking, liquor has a close connection with Chinese culture in both ancient and modern times. Spirits culture has been playing a quite important role in Chinese people's life for a long time. Our ancestors used liquor to enjoy themselves while writing poetry, or to make a toast to their relatives and friends during a feast. Spirits culture, as a kind of culture form, is also an inseparable part in the life of ordinary Chinese people in occasions like birthday parties, farewell dinners, weddings, etc.

译文

五粮液酒文化与茅台酒文化

　　五粮液与茅台在白酒品牌中都享有有很高的地位和知名度，这两种白酒都是采天地之灵气，融谷物之精华而成，是白酒品牌中成大器者。从这两个具有代表性的名酒中，我们不仅可以享受美酒，更可以品到中华民族几千年的文化传统。这两种酒的文化都起源于《易经》中的贵和持中思想。

　　五粮液以儒家文化的中庸文化为其文化品质，即追求传统与现代的和谐统一。从过去到现在，五粮液一直以其独特的自然生态环境，独有的638年明代古窖、独有的五粮配方和酿造工艺，不断地发展和壮大五粮液文化。五粮液注重感官上的色、香、味俱全，讲究酒的内在品质，注重饮用后的绵长回味。它的性情柔中有刚，味觉层次全面丰富，谐调地调动了人的视觉、嗅觉和味觉，恰到好处地达到了真正的中庸和谐。五粮液文化除了传承了儒家的中庸文化，还在历史长河中演变为刚健有为、积极进取的文化。这可以从五粮液世纪广场的"强鱼吃弱鱼"文化雕塑中看出。人们一进五粮液集团所在地就感受到市场竞争的残酷和惨烈，明白市场竞争的强者生存、弱者灭亡的生存法则，因而五粮液只能不断地做成强中之强，才有生存和发展的空间。

　　茅台酒文化逐渐地演变成道家的无为而治的文化品质，更加重视顺其自然之理。茅台商标的蓝色可以让人联想到大海，以及可以容纳和承接宇宙的气魄。从茅台镇的图腾可以看出，茅台崇尚一种静态与平和。茅台以道家无为文化为其文化品质。开放式发酵与封闭式发酵对应太极中的"一阴一阳之谓道"的理念。在多年的市场竞争中，茅台长期高举"国酒"的旗帜，以此号召中国白酒市场。茅台多年来致力于酒类品质的坚持，力倡健康品质，形成无可争锋的态势。茅台希求在激烈的市场竞争中成为无与匹敌的"酒业帝王"，它除了坚持酒的好品质，还一直坚持自己的国酒文化。其文化品性主要体现在"忠孝节义"四个字上。为国争光，诚于国事；谓之忠。儿遂母愿，殷勤于家；谓之孝。不羡繁华，不易其地；谓之节。护身健体，不伤饮者；谓之义。忠孝节义四全，是谓国酒文化。

　　五粮液和茅台虽然文化品质有差异，但是都获得了广大消费者的认同和喜爱，两个品牌都在传承中国的传统文化并且将之复兴。今天，五粮液和茅台将走出国门，走向世界。要想使"中国名酒"成为"世界名酒"必将加大对中国传统文化的介绍，使中国传统文化能被西方社会和人们所接受。

Unit 3

Section A

I.

1 The word "alcohol" is *derived* from the Arabic word "al Kuhul" which originally referred to a powder used by women for make-up.

2 Throughout China's ancient and modern drinks, the cultural effects of Baijiu are remarkable: when one feels happy, he always drinks "grape wine with *luminous jade* cup"; when *decadent*, "Enjoy while one can"; when missing relatives and friends, "How long will the full moon appear? Wine cup in hand, I ask the sky."; when gathering with friends, "When one drinks with a good friend, a thousand cups are not enough"; when feeling lonely, "I raise my cup"; when saying goodbye, "I invite you to drink a cup of wine again; West of the Sunny Pass no friends will be seen."

3 Baijiu adds a lot of topics to the dinner, for instance, while people are drinking and talking because of the atmosphere of Baijiu, they are cheerfully chatting and laughing, which can warm the whole body. If you are in a good mood, you will naturally be healthier.

4 Baijiu can be used in medicine because Baijiu is a good *solvent*. It can dissolve many poorly soluble or even indissoluble substance.

5 Red grape wine has a polyphenolic chemical called *anthocyanidin*, which is the source of red and has a strong *antioxidant* function.

6 It has the effect of skin-moisturizing and blood-activating, preferred by people who wish to stay young and beautiful .

7 After the bath, the skin is as white as jade and the body is warm all over.

8 Millet Baijiu is the most ideal cooking liquor, because it contains moderate ethanol content between beer and Baijiu, and it is rich in amino acids, which can produce *amino acid sodium salt* in the cooking process with salt, namely *monosodium glutamate*, which

can increase the delicacy of dishes.

III.

1 j   2 f   3 g   4 i   5 h   6 b   7 c   8 e   9 d   10 a

IV.

1 permeate   2 derive   3 decoct   4 dispersed   5 stimulus
6 therapeutic   7 decadent   8 cosmetic   9 unblock   10 Nocturnal

V.

Baijiu and its drinking culture have always played a significant role in Chinese history. In the Song Dynasty, Baijiu became the main beverage Chinese people took. Chinese Baijiu has a complicated production process and can be made from various raw materials, making it one of the six world-famous distilled liquor. There are lots of excellent brands in China favored by different groups of people. In today's society, the drinking culture has undergone unprecedented enrichment and development. The drinking customs and rituals in different regions and on different occasions have become an important part of Chinese people's daily life. In thousands of years of civilization, Baijiu has penetrated into almost every aspect of social life, such as literary creation, diet and health care.

译文

饮酒对健康的好处

酒精的语源来自阿拉伯语的"al kuhul",原本是妇女化妆用的一种粉末,后指"酒之精华",谓酒精。一般的酒,除含乙醇外,尚含酯类、酸类、酚类及氨基酸等物质,加之多是由五谷杂粮、果实制成,酒有水谷之气,味辛甘、性热,易入心肝二经,所以有通畅血脉、行气活血、祛风散寒、清除冷积、医治胃寒、强健脾胃以及行药势的功效。适量饮酒,可使人思维活跃,激发人的智慧,尚可强心提神、消除疲劳、促进睡眠。酒进入体内,可扩张血管,增加血流量。酒对味觉、嗅觉是一种刺激，可反射性地增加呼吸量、增进食欲。经测试得知，人体内少量酒精，可以提高血液中的高密度脂蛋白的含量和降低低密度脂蛋白的水平，为此少量饮酒可减少因脂肪沉积引起的血管硬化、阻塞的机会。适量饮酒可以促进血液循环，能够减少死于心脏病的可能性。酒精其实有助于防止血栓的形成，因而有效地防止了中风和心脏病发作的可能，适量饮酒的人比滴酒不沾的人还长寿！

1. 酒可载情

酒使人精神镇静、畅快，这是酒自古以来能流传至今的一种精神力量。纵观中华古今饮品，酒所起的文化功效甚为显著：高兴时"葡萄美酒夜光杯"；颓废时"今朝有酒今朝醉"；怀念亲友时"明月几时有，把酒问青天"；与友人会聚一堂时"酒逢知己千杯少"；孤独时"举杯邀明月，对影成三人"；惜别时"劝君更进一杯酒，西出阳关无故人"……可说助兴者酒，浇愁者亦酒，酒渗透中国人社会生活的各个角落，成为一种文化的载体，被人们誉为"酒文化"，为人类文化生活增加众多色彩光辉。俗语说"无酒不成席"，酒给席间增添很多话题，边饮边侃，融融浓情和酒一起暖遍全身，酒兴所致，心扉敞开，欢声笑语，笼罩席间。心情舒畅，自然身体会更加健康，当然如沉湎滥饮不仅不解忧，反而影响健康。

2. 酒可入药

酒不仅可载情，尚可治病、滋补。酒是"救人的良药"，但有时也是"杀人之利器"，鸩酒一类的毒酒便可治人于死地。酒可入药是因为酒精是一种很好的溶剂，它可溶解许多难溶甚至不溶于水的物质，用它来泡制药酒，有的比水煎中药疗效好。而且药酒进入体内被吸收后立即进入血液，能更好发挥药性，从而起到治疗滋补之功效。为此，中医常有处方让患者用酒冲服，或煎药时使用药引。酒不仅可内服，而且能用于外科。除酒精消毒外，酒可以涂于患处，治疗跌打扭伤、关节炎、神经麻木等，如虎骨酒、史国公酒等。东汉名医华佗，曾用酒冲泡叫作"麻沸散"的药，作为麻醉剂完成外科手术，并获得成功。此为公元2世纪的事情，系世界首创，虽不再使用，但在未发明麻醉剂的时代，酒可起到了救助外科病患者的作用。不同的酒有不同医疗作用，最新的研究成果说，每天饮一点烈性酒，可减少冠心病发作，从而减少冠心病引起死亡的危险性。近年来红葡萄酒在中国很畅销，备受青睐。因为适量饮用葡萄酒不仅可防衰老，而且尚可预防因机体老化引发的有关疾病。红葡萄酒中有一种属于多酚的化学物质叫花色素（anthocyanidin），是红色的来源，它有很强的抗氧化的功能。若能抗氧化，便能抗衰老。花色素主要存在于红葡萄的皮和籽中，在酿制过程时，转变为抗氧化作用的物质。

3. 酒与健美

酒有健美之功效早在唐代苏敬等人所著的《新修本草》一书中已有记述："暖腰肾、驻颜色、耐寒"。这里是指葡萄酒，在7世纪中叶，葡萄酒传入中国并在中国得到发展。还有桃花酒，是将三月新采的桃花阴干后浸泡在上等酒中，贮15日便为桃花酒，饮用该酒，有润肤、活血的功效，使人青春美容长驻。白鸽煮酒、龙眼和气酒也有美容作用。为使毛发肌肤健美，中国古代就有用酒洗浴的做法。日本盛行一种酒浴。入浴前，将0.75kg的"玉之肤"饮浴两用酒加入浴池水中，洗浴后皮肤洁白如玉，周身暖和。"玉之肤"浴酒是把发酵酒糟和米酒混和，再经蒸制而成，是清酒的一种。后经医学专家研究，酒对皮肤有良性刺激，能加速血液循环，对身体大有裨益。为此，此后日本流行酒浴。

4. 酒与烹饪

在烹饪美味菜肴时，适量用酒，能去腥起香，使菜肴香甜可口。因为酒的主要成分是乙醇，沸点较低，一经加热，很易挥发，便把鱼、肉等动物的腥膻怪味带走。烹饪用酒最理想的是黄酒，因为它含乙醇量适中，介于啤酒和白酒之间，而且黄酒中富含氨基酸，在烹饪中与盐生成氨基酸纳盐，即味精，能增加菜肴的鲜味。加之黄酒的酒药中配有芳香的中药材，用它作料酒，菜肴会有一种特殊的香味。当然，在无黄酒的情况下，其他酒也可以用。

Section B

I.

1 O 2 H 3 J 4 D 5 G 6 I 7 N 8 K 9 E 10 M

II.

Baijiu, as a special form of Chinese culture, has a history of more than 5,000 years. According to the book *The Spring and Autumn in the Cup* by Lin Chao, drinking is something of learning rather than eating and drinking. There are many stories about Baijiu in Chinese history. The great poet Li Bai in the Tang Dynasty could "write hundreds of poems after drinking white spirit", and the more Baijiu he drank, the better his poem would be. Baijiu plays an extremely important role in Chinese folk custom. Spirits are used to celebrate different festivals, wedding ceremonies and birthday parties, to memorize the departed, to welcome and send off relatives and friends, to congratulate the good news and to get rid of anxiety, to cure diseases and prolong life, both for emperors and ordinary people.

译文

如何健康饮酒

　　酒是人们生活中常见的一种饮料，多饮是危害健康的，但是要想让喝酒达到养生的目的，那么就要知道如何健康饮酒，并且明白喝酒不能吃什么。本文从如何健康饮酒、吃什么喝酒不容易醉等方面进行详细介绍。

　　我们在当今社会中，常可以听到有人因喝酒而中毒死亡的消息，因此我们劝诫喝酒的人选择健康饮酒的方式。下面就告诉大家如何健康饮酒。

六个健康喝酒的方式

1. 在酒里加点苏打

　　在酒里加点苏打可以降低人体对酒精的吸收，这样比较不容易醉，自然脸就不容易红了。

2. 空腹时不要饮酒

　　要选择一面饮酒，一面进食，因为酒在胃内停留的时间长，酒精受胃酸的干扰，吸收缓慢，这样就不容易酒醉。酒的特征之一就是在消化器官中的吸收非常快。进入体内的酒精，20%被胃吸收、80%被小肠吸收，可以溶入血液中运往身体的各角落。

　　如果胃中有食物的话，酒精被移往小肠的速度就会减缓，而如果是空腹没有任何东西覆在胃黏膜上的话，酒精便会在胃中畅通无阻，一路直奔小肠。吸收速度加快，不久血液中酒精浓度便急剧上升。

　　就是说，如果连干几杯或空腹饮酒，一瞬间体内血液中的酒精浓度就会升得很高，令自己进入危险的麻痹状态。所以，这种饮酒法要尽量避免。

3. 不要大口猛喝

要慢慢喝酒，不时地停顿一下，喝酒时不要喝碳酸饮料，如可乐、汽水等，以免加快身体吸收酒精的速度。

4. 不要用错醒酒物

人们一直认为茶可以解酒，甚至会喝雪碧或可乐来解酒。然而，解酒的最佳选择其实是果汁，特别是橙汁能起到很好的解酒作用，因为果汁中含有果糖，可以有效帮助酒精燃烧。

5. 不要多种酒混合饮

因为各种酒成分、含量不同，互相混杂，会起变化，使人易醉或饮后不适，甚至头痛。

6. 饮酒后切不要洗澡

人饮酒后体内贮存的葡萄糖在洗澡时会被体力活动消耗掉，引起血糖含量减少，体温急剧下降，而酒精抑制了肝脏正常的活动，阻碍体内葡萄糖贮存的恢复，以致危及生命，引起死亡。

喝酒的错误观点

1. 喝酒吃肉不吃饭

这是慢性自杀，白酒里没什么营养，只有热量。常这样不吃主食，不仅致消化不良，也易致营养不良。光吃肉不吃主食，一会导致营养不良，因为人的营养70%来自主食，其他来自副食。二会导致大量缺乏维生素B（VB），身体缺了VB，酶就会缺乏活性，身体很多功能就难以正常发挥，而主食里含有大量的VB。

2. 饮酒并不有助于睡眠

很多人以为喝酒就会使人昏昏欲睡，其实并非如此。少量饮酒可以使人兴奋，大量喝后则转为抑制，呼吸较浅弱，代谢降低，体温下降。

饮酒到一定程度，身体很多部位都会充血，比如咽喉部充血了，睡时就会打呼噜，肌肉松弛了或塌下了，就会使呼吸不顺畅或致呼吸道阻塞，易引发心脑血管疾病或意外死亡。

Unit 4

Section A

I.

1. Chuan-Qian (Sichuan province and Guizhou province) production area, Huang-Huai (the Yellow River and the Huai River valleys) production area and lakes (Hunan province and Hubei province) production area.
2. Water, soil, climate, temperature and microbe
3. As early as the pre-Qin period, Bo people's technology of making betel pepper (the origin of Baijiu) had been matured. In the late 3,000 years, the prescription of Baijiu and the process of making Baijiu has been innovating and perfecting unremittingly.
4. This region accumulated swarms of Baijiu brands and series of Baijiu products, best representing the rich deposits of Chinese Baijiu culture.
5. It will drive the products' sustained innovation, increase brands awareness and boost sales effectively.
6. Abundant water resources and Baijiu-making raw materials.
7. It is the meeting point of Luzhou-flavor Baijiu belt from sichuan to Anhui, Jiangsu provinces and Moutai-Luzhou-flavor Baijiu belt from shanxi province to Guangdong and Guangxi regions.
8. The tradition of brewing Baijiu maintains from ancient times to the present all around China. Also, distinctive region characteristics construct a basis for unique Chinese Baijiu culture.

II.
1 c   2 h   3 g   4 j   5 e   6 i   7 d   8 a   9 f   10 b

III.
1 differentiate   2 innovate   3 sustainable   4 endowed

5 endeavor   6 forge   7 domain   8 complement
9 distill   10 foster

V.

2010年9月19日,总投资约50亿元、占地250多公顷的"中国白酒金三角——酒都宜宾·五粮液文化特色街区"盛大启动。五粮液文化特色街区是迄今为止四川省投资最大的以酒文化为主题的特色街区。五粮液文化特色街区项目分为南北两大片区,十里酒城生态园的生态特色,旧州塔公园将充分展示川酒特色;"利川永"等酿酒古窖池将被用作酒都历史古迹进行包装和保护,规划和修缮为宜宾"酒之源"。并通过规划建设白酒文化博物馆、流杯池白酒文化主题公园等项目,将集中展现以酒都宜宾为代表的川酒四千年酿造历史。

译文

一方水土养一方人。同样,一个地区环境的特点决定了当地酒的独特性,一方水土也能酝酿出有别于他处的美酒。可以说,世界上每一种美酒,都是地方独有的特产,中国白酒更是如此。在中国,白酒是一种历史悠久的烈酒,已有3000多年的历史。酒是风物志,白酒的酿造高度依赖于"水(质)、土(壤)、气(温)、气(候)、生(微生物)"的自然环境,不同的地理环境与风土人情赋予了每一个地域出产的白酒截然不同的属性,即是通常所说的产区特色。根据影响白酒生产的地理因素和生产工艺,以及现有的产能产量来看,中国白酒版图可以划分成川黔产区(四川和贵州),黄淮产区(黄河和淮河)和两湖产区(湖南省和湖北省)等三大白酒产区。

川黔产区

举世公认的是,作为古老而神秘的僰族祖地,古僰人繁衍生息的川黔地区是浓香型和酱香型世界顶级白酒的发源地。也是中国白酒文化的缩影。

早在先秦时期,僰人酿蒟酱(一种果酒)达到了技术成熟期。近3000年来,白酒的配方和制作工艺不断创新和完善。为了保持白酒原有的生态文化和白酒酿造业的持续健康发展,长江流域、岷江河流域和赤水河流域的大部分酿酒企业仍坚持传统的白酒酿造技艺。酒城、酒庄、酒楼星罗棋布,汇聚了大量的白酒品牌和白酒系列产品,最能代表中国白酒文化的丰富底蕴。它已成为打造中国世界级区域白酒品牌的核心。

川黔产区集聚了大量中国传统白酒品牌。由宜宾、泸州和遵义等三市所构成的"中国白酒金三角",白酒产量占全国20%左右,孕育形成了享誉全球的五粮液、茅台、剑南春、沱牌、水井坊、泸州老窖、郎酒等国际品牌和等中国最著名白酒品牌,扛起了中国白酒产业的半壁江山,在行业内部素有"最大的产业集群、最大的品牌群、最大的产能群、最好的政策洼地"之称,被联合国粮农组织誉为"地球同纬度上最适合酿造优质纯正蒸馏酒的生态区"。

黄淮产区

黄淮流域是中华文明的发祥地之一,名流雅士层出不穷。自秦汉以来,黄淮流域的社会、经济、文化空前繁荣。特别是从三国(公元220-265年)到唐宋时期(公元960-1279年),黄淮地区成为中华文明传承的核心区域。

黄淮产区位于长江流域和黄河流域之间,不仅受黄淮两大水系的影响,而且南北交界的地理位置使其拥有极其独特的温带季风气候气候和地理环境。独特的气候,巨大的粮食生产量和无法复制的中国南北方交汇点的地理位置使它成为酿造浓香淡雅型白酒的宝地。符合自己地域特色风格和人文特点的黄淮名酒,包括江苏洋河、安徽古井、河南杜康宋河和山东孔子家族酒。这些名酒不仅与黄淮文明的历史、地域相辅相成,相映成辉,也使得中国黄淮名酒带一直不同凡响。

2007年,豫鲁苏皖四省提出白酒抱团性战略,特别是四省白酒风格的确定,从重香到重味的转变,使得黄淮产区成为中国白酒领域的一支重要力量。2015年,黄淮著名白酒发展联盟诞生,旨在将品牌做大做强,打造统一口味和风格,提升品牌影响力,开拓市场。联盟的成立将推动产品的持续创新,提高品牌知名度,有效促进销售。

两湖产区

两湖地区位于长江中游,有史以来就是我国农业和经济发达之地,素有"鱼米之乡"之称。该产区因拥有优质的酿酒作物和水资源禀赋而成为中国白酒产区中的重要一级。

从四川到安徽、江苏的长江流域的白酒浓香带以及从陕西到两广地区的白酒兼香带在荆州、宜昌、常德地区交汇。得天独厚的水土气候和微生物条件使得该地区白酒香型丰富,酱香,清香,浓香,馥郁香,特香等香型百花争艳。同时,该产区白酒产量居全国前列,产业集群度高,销售业绩优异,成为继川黔产区、黄淮产区之后的第三大白酒产区。近年来，两湖产区出产的名酒如湖北稻花香、枝江，湖南浏阳河、酒鬼酒等品牌，知名度和销售业绩令人瞩目。

Section B

I.
1 B   2 C   3 A   4 D   5 A   6 B   7 D   8 C

II.

Liquor was closely related to literature and art in the Tang and Song Dynasties, which can be seen from literature and art works such as Tang Poetry and Song Ci. This phenomenon makes the Tang and Song Dynasties a special period in the history of Chinese spirits culture. Ming Dynasty was a period of great development of liquor industry, the variety and yield of liquor greatly exceeded any previous dynasties. By the Qing Dynasty, big liquor shops appeared one after another, the output increased year by year, and the sales expanded constantly. In order to expand and facilitate the sale, big liquor shops unified the varieties, the specification and the packing form of liquor, and opened the tavern or the winery to manage the whole sale and retail business.

译文

中国的波尔多：长江上游的"中国白酒金三角"

在中国西南，四川和贵州的交界处，宜宾、泸州和遵义三座城市孕育了中国标志性的白酒品牌，如茅台、五粮液、郎酒和泸州老窖。现在，这一地区正在创建能与法国波尔多红酒相媲美的"中国白酒金三角"。

好水酿好酒

虽然粮食酒，或者说白酒在全中国范围内均有出产，但白酒金三角地区集合了全中国最多的白酒生产设备及生产厂家。

地处亚热带气候的四川盆地，这一地区湿润的天气、清洁的水源、肥沃的土壤和盆地地形为酿制以高粱、小麦或大米为原料的优质白酒创造了得天独厚的条件，形成了历史悠久的白酒文化和酿酒传统。

正如俗语所说："好水之地出好酒"。沱江、赤水河、岷江和长江横贯这一地区，丰富而清洁的水资源为美酒的酿制打下了坚实的基础。

四川省宜宾五粮液集团有限公司副总经理彭智辅说："这一地区的自然环境是一个特殊的生态系统，因其成因复杂而无法复制，无可比拟。"

"没有赤水河流域健康的生态系统就没有茅台酒。赤水河是茅台酒的生命河。"贵州茅台集团董事长、总经理李保芳说。赤水河流域茅台地区细菌和古菌多样性以及赤水河水滋养的酿酒作物孕育了茅台酒的独特风味。

同时，孕育美酒的珍贵生态系统也是这些企业与当地政府共同努力保护的结果。

为了保护赤水河沿岸的生态，茅台集团已经投入4.68亿元，建立了5个污水处理厂，年处理污水能力达到200万吨。

从2006年起，泸州开始建立"中国白酒金三角"白酒产业园，并力邀白酒产业相关上下游企业入驻。产业集群在高粱种植、酿酒、白酒贮藏、包装、供应、物流配送等方面进行了联合规范的改进。同时，通过在产业园提供污水处理厂、国家标准质量监督中心等基础设施和配套服务，也保证了整个产业链的环保。

为实现优质南方糯红高粱（一种特殊的酿酒作物）的当地种植，目前主要酿酒企业如著名的泸州白酒生产商泸州老窖股份有限公司，已经开始建立自己的有机原料基地。此举不仅有助于保护白酒产品质量，也有助于促进当地农业产业发展，帮助农民增加收入。

白酒文化，另一张名片

长江流域的酿酒史可以追溯到千年前的唐朝（公元618-907年），而中国的饮酒史则可以追溯到商朝（公元前1600-1046年）。

白酒是中国最著名的烈酒，几乎是中国所有的节日场合必不可少的饮品。从婚宴到商务宴会，白酒都是人们的首选。许多中国古代诗人，如李白和杜甫，他们的传奇人生和经典诗作都与白酒密不可分。

独特的白酒文化使这些酿酒企业的酿造工艺和酿酒原料成为到该地区游览的游客的必游之地。这些酿造工艺被列为国家非物质文化遗产。

泸州老窖是"金三角"地区历史最悠久的酿酒企业之一，其最大的魅力在于其大名鼎鼎的酒窖，其历史可以追溯到1537年。

利用起源于明朝（公元1368-1644年）和清朝（公元1644-1911年）的酿造工艺，该品牌继续以同样的方式酿造了白酒并传承23代。

"酒窖历史越悠久，酿出来的酒越好。"泸州老窖第23代窖藏白酒品鉴传承人曾娜说。

截至2013年，泸州拥有1615年的酿酒酒窖、16个传承自明清时期的白酒作坊、3个天然酒窟，成为体验白酒文化的独一无二的旅游目的地。

白酒正走向世界

尽管白酒是世界上喝得最多的烈酒，但它只占全球市场不到10%的份额，而且很少在鸡尾酒菜单上和其他烈酒一起出现。

为了吸引更多的消费者，尤其是海外市场的消费者，中国的白酒生产商除了延续古老的酿酒传统外，还在引进新技术和创新产品。

在五粮液的故乡宜宾，已经建立了一所专门为该行业培养人才的学院，而在泸州老窖，像曾娜这样的品酒大师正在开发烈酒和果汁混合的鸡尾酒，以迎合初学者的口味。

近年来，泸州老窖一直致力于振兴中华优秀传统文化，促进传统文化在全球范围内的传播。

2017年，泸州老窖举办了"国窖1573""让世界感受中国的味道"国际诗歌文化大会等一系列全球慈善活动，并积极参与国家活动，在世界舞台上弘扬中华优秀传统文化。

泸州老窖总经理林峰表示："外国市场仍对中国白酒生产商开放，中国白酒在未来5到10年将在全球市场取得突破。"

Unit 5

Section A

I.

1 There are eight major steps of the production process of Baijiu: (1) ingredient formulation; (2) grinding and cooking; (3) mixing and cooling; (4) mixing with daqu; (5) loading to the fermentation vessel; (6) alcoholic fermentation; (7) distillation; and (8) aging.

2 It has been shown that a lower steam flow rate may not provide the *thermal* condition for the complete *evaporation* of *ethanol*, while a higher steam flow rate may cause the fermented grains to stick together and increase the *diffusion* resistance.

3 Aging plays an essential role in the flavor of liquors, since a variety of aromatic compounds (mainly acids and esters) are balanced during this process through physical changes.

4 High temperature is used to obtain daqu dominated by thermophilic bacteria, stacking the fermented material for a few days (2–4 days) before alcoholic fermentation with the purpose of increasing the number of microorganisms, particularly yeasts, and balancing the chemical components.

5 Experience has shown that the pit not only provides the place for fermentation but also contributes to the flavor of Baijiu.

6 Back-slopping technique is introduced by adding the fermentation residue during the beginning of alcoholic fermentation for a better regulation of the fermentation process.

7 The fermentation time is dependent on various factors such as climate and moisture content. Alcoholic fermentation takes about 1 month.

8 Due to the subtropical monsoon climate, alcoholic fermentation takes from 60 to 90 days for the Baijiu produced from southern China. Compared to southern China, northern China has the

characteristics of low moisture and long daylight. Therefore, alcoholic fermentation takes approximately 45 to 60 days.

II.

a-5   b-4   c-8   d-9   e-6   f-10   g-2   h-3   i-7   j-1

III.

1 circulation   2 additive   3 distillation   4 barley
5 formulate   6 diffusion 7 fermentation   8 aging
9 texture   10 aromatic

IV.

Chinese people began to make spirits with grain seven thousand years ago. Generally speaking, spirits has a close connection with culture in China in both ancient and modern times. Chinese spirits culture has been playing a quite important role in Chinese people's life for a long time. Our Chinese ancestors used spirits to enjoy themselves while writing poetry, or to make a toast to their relatives and friends during a feast. spirits culture, as a kind of culture form, is also an inseparable part in the life of ordinary Chinese people such as birthday party, farewell dinner, wedding, etc.

译文

白酒的生产

白酒生产技术是中国珍贵的民族遗产，现代白酒是技术进步渐进的结果。中国传统的发酵技术已使中国人受益数百年。传统发酵产品改善了中国人的日常饮食习惯，丰富了世界餐饮文化。中国白酒被认为是中国人发酵产品的骨干，因为其产量增长迅速，目前每年超过1200万吨。

尽管有超过1万家工厂使用自己的技术生产白酒，但白酒生产的原则仍然相同。总体而言，白酒的生产涉及八个主要步骤：（1）配料配方（2）研磨和烹饪（3）混合和冷却（4）与大曲混合（5）装载到发酵容器（6）酒精发酵（7）蒸馏和（8）

陈化。

　　以高粱为主要原料。研磨高粱以释放淀粉，目的是增加烹饪和微生物相关区域并获得理想的质量凝聚力。这一步对白酒的品质起着重要的作用，因为软磨会导致无效的糖化，苛刻的磨碎会影响白酒的风味。烹饪的目的是引入淀粉进行糊化。混合和冷却：将温度高于85℃的水和其他添加剂混合，以获得均匀的质地和理想的风味。冷却用于降低温度以准备与活性微生物群（大曲）混合。当发酵物料（大曲粒）的温度降至18℃至20℃之间时，将混合物装入瓦罐中。酒精发酵此步骤通常在瓦罐中进行。该罐子从一批到另一批循环使用。发酵时间取决于各种因素，如气候和含水量。酒精发酵需要约1个月。

　　蒸馏是白酒风味发展的关键一步。蒸馏效率取决于蒸汽流量，含水量，蒸馏速度和材料的孔隙度。已经表明，较低的蒸汽流量可能无法为乙醇的完全蒸发提供热条件，而较高的蒸汽流量可能会使发酵的谷物粘在一起并增加扩散阻力。陈化对白酒口味极其重要，由于各种芳香化合物（主要是酸和酯）在这一过程中通过物理变化相互作用和化学反应。一般来说，酱香型白酒的陈化时间超过3年，而强，弱香型白酒至少需要1年。

　　使用高温堆积发酵，这是生产酱香味白酒的关键过程。在大曲制作过程中，通过高温来获得由嗜热细菌占主导地位的大曲。然而，酵母菌和霉菌在酒精发酵中起着重要作用，因为它们将可发酵糖转化为酒精和芳香化合物。因此，高粱添加大曲后直接引入高温堆积发酵。这种技术的原理相当简单，就是在发酵前将发酵材料堆积几天（2-4天），目的是增加微生物，特别是酵母的数量，平衡化学成分。

　　窖池的使用：窖池是由泥土制成的地窖，泥浆坑的平均体积为6立方米至8立方米。当发酵物料（大曲谷物）的温度降至20℃至21℃时，将混合物装入窖池中。经验表明，该池不仅提供了发酵的场所，而且还有助于白酒的风味。

　　使用续槽技术：白酒通过酒精发酵和蒸馏获得，产品通常不稳定。通过在酒精发酵开始阶段添加发酵残渣以更好地调节发酵过程而引入续槽技术。该技术可以生成微生物群落的最佳组成，这反过来增加了天然发酵的成功。与其他类型的白酒生产相比，该技术被认为是重要的一步。

　　中国南方生产的白酒，如五粮液酒和剑南春酒，主要以高粱，大米，糯米，小麦，玉米为原料，而中国北方生产的白酒则以高粱为原料唯一的材料。中国南方生产的浓香型白酒以小麦为原料制作大曲，而华北地区则以大麦，小麦和豌豆为原料制作大曲。在与高粱和其他谷物混合之前，大曲通常制成块状并磨碎。由于亚热带季风气候，中国南方生产的白酒酒精发酵需要60至90天。与华南地区相比，华北地区具有湿度低，日照时间长的特点。因此，酒精发酵大约需要45到60天。总之，每种白酒都有其特有的关键技术，导致白酒的口味不同。它们都与特定的制作过程和酒精发酵过程有关。每个步骤的变化都会改变微生物分布，从而改变微生物代谢产物。因此，可以生产不同口味的白酒。

Section B

I.

1 G　2 B　3 F　4 A　5 H　6 K　7 N　8 M　9 L　10 J

II.

China is the origin of rice culture in the world, which is rich and sufficient in raw materials of liquor making. China has a long history of making and drinking liquor. Pottery excavated proves that Chinese people started brewing mild grain liquor 5,000 years ago. Many evidence, including Chinese inscriptions on bones or tortoise shells, bronze inscription of the Chinese character "jiu" and the rich varieties of bronze wine vessels, show that the liquor industry was already well developed during the Shang and Zhou Dynasties. The invention of the distilling process was a milestone for the Chinese liquor industry. About 1,000 years ago, in the Song Dynasty (960-1279), the Chinese were able to make highly alcoholic grain liquor using the distilling process, which is now called Baijiu.

译文

白酒分类

中国不同地区的各种工艺生产出数百种不同类型的白酒。白酒可以根据生产技术（固态和半固态），发酵剂类型（大曲、小曲和麸曲）以及产品风味（酱、浓、清等）进行分类。不同类型的白酒由于其独特的生产技术而具有不同的味道。

根据制造技术分类

固态发酵白酒是一种将微生物培养物在没有液体（水相）的情况下在固体基质上生长的过程。这种方法已被广泛用于大多数著名的白酒生产技术，并且历史悠久，并且已经传承了几代人。这种白酒通常由高粱，小麦，大米，糯米和玉米等谷物通过复杂的SSF工艺生产，包括（1）材料制备、（2）打曲、（3）固态发酵、（4）固态蒸馏和（5）熟化。这种技术导致含有约60%水的发酵材料。根据不同的发酵工艺和操作条件，它可以生产不同口味的组分。因此，SSF被认为是可以产生最大种类的白酒的一种方法，每种产品都具有不同的风味和特征。

半固态发酵白酒，由桂林三花酒和泉州香山酒代表。发酵过程在半固态下运行。

液态发酵白酒，以红星二锅头酒为代表。所有生产过程包括糖化，发酵和液态下的蒸馏。

根据所使用的发酵剂，可区分出三种不同类型的白酒。

以大曲为发酵剂的白酒以四种最出名的白酒代表：（茅台酒，五粮液酒，汾酒和泸州老窖酒）。大曲是一种由原麦，大麦和/或豌豆制成的谷物曲。将浸湿的材料转移到模压机中并成形为砖块，每块重约1.5kg至4.5kg，其中平面或一侧呈凸形。因为它的体积很大，因此被命名为大曲（大发酵剂）。一般来说，大曲中存在四类微生物（细菌、酵母、丝状真菌和放线菌）。使用大曲生产的白酒口味丰富，发酵时间长，酒精产量低。

以小曲为首发的白酒，以桂林三花酒和浏阳河酒为代表。与大曲相比，小曲是一种小米，由米或米糠制成。与大曲不同，小曲中只有少数几种微生物存在，包括根霉菌，毛霉菌，乳酸菌和酵母菌。这些微生物主要是良好的发酵表演者；因此，在生产这种白酒时使用少量的发酵剂和短时间的发酵时间。然而，由于涉及微生物的几种类型，使用小曲生产的白酒含有比使用大曲生产的更少的风味。

以二锅头酒为代表的使用（麸曲）作为发酵剂制作的白酒。麸曲与其他两种发酵剂不同，由麸制成，仅含有纯曲霉菌培养物。曲霉是一种众所周知的淀粉降解剂，可将淀粉转化为可发酵的糖。生产的白酒具有风味清淡，酒产量高的特点。

白酒口味根据其风味，白酒可分为三大类。

酱香型（茅香型）白酒，茅台酒和郎酒等酱香型白酒，具有酱香味，持久香气的风味。主要代表性的芳香化合物是酚类化合物，丁香酸，以及少量的氨基酸，酸和酯。

浓香型（泸香型）白酒，具有香味浓郁，口感柔和，回味无穷，以泸州老窖和五粮液等浓香味的白酒为代表。代表性的芳香化合物主要是己酸乙酯，与乳酸乙酯，乙酸乙酯和丁酸乙酯协调平衡。

清香型（汾香型）白酒，汾酒和二锅头酒等清香型白酒，口感纯正温和，醇厚甜美，回味清爽。主要的香气化合物是乙酸乙酯，其中含有相当量的乳酸乙酯。

这三种风味类型的特点非常典型且具有代表性，它们占中国白酒的约60%至70%。除了这三种类型之外，还有使用不同的技术生产具有特定风味和香味特征的西凤酒、董酒、四特酒、衡水老白干酒、酒鬼酒等白酒。随着科学技术的发展和各种起始原料的使用，可生产各种口味的白酒。未来还需要更严格和准确的标准，以基于其风味丰富的化合物来区分白酒。

其他分类白酒还可分为高酒度（酒精含量>50%v/v，每100毫升饮料酒中含有乙醇的毫升数），中等酒精含量（41%-50%v/v）和低酒精含量（<40%v/v）；根据发酵原料，有高粱酒，玉米酒和水稻酒等。

Unit 6

Section A

I.

a Observing, Smelling, Tasting, Evaluating
b strong-aroma, sauce-aroma, light-aroma, rice-aroma, Chi-aroma, Feng-aroma, mixed-aroma, Laobaigan-aroma, sesame-aroma, Te-aroma, herbal-aroma, extra-strong aroma

II.

a-6   b-8   c-1   d-5   e-10   f-2   g-9   h-3   i-4   j-7

III.

1 overstrain   2 nostril   3 coordinate   4 sensitivity
5 turbid   6 sediment   7 mellow   8 mediocre
9 aftertaste   10 poignant

IV.

　　在探讨白酒时，人们都以"白酒"为统称，于是人们的心中形成了白酒种类仅仅只有一种的误解。而事实上，白酒就如一个枝繁叶茂的大树，分支繁多。根据不同的原料，生产方式和口味，白酒可分为四种基本香型——浓香型，清香型，酱香型，米香型——和一系列其他更细、基于品牌特征的香型。现代白酒分类体系为现代的一种发明，虽有局限性，但对于品鉴者来说有很实用的参考价值。过去，人们总是给特定的白酒种类打上地方标签。而近来数十年，地理界线不再泾渭分明，新的白酒香型层出不穷。

译文

享受白酒品鉴的乐趣

　　白酒的风格多样，只要品酒人事先做好准备，进行系统的品鉴训练，白酒就会回报品酒人丰富的品酒乐趣。在社交之中，品酒人就多了一个扩大交际深化友谊的工具。

品鉴前的准备

品鉴场地： 品酒环境会直接影响品鉴的结果，选择一个合适的品酒场地是极其重要的。首先，品鉴场地应该空气清新，不允许有烟雾及异样气味存在。其次，品酒场地的光线应该自然，不适合有色光源的存在。如酒吧不是一个合适的品鉴场所，酒吧的气味和灯光都不利于品鉴的进行。另外，品鉴场地中应该设有洗漱水池以便及时清理酒杯及其他器具。同时，品鉴场地里的温度不宜过高或过低，过高或过低的温度都会影响酒香的发散，室内温度以18-22℃为宜。

品酒杯： 酒杯的大小、色泽、形状、质量和容量都会对品鉴的结果产生影响。所以选择合适的酒杯，对得到准确的品鉴结果至关重要。标准的白酒品鉴酒杯应该为无色透明、无花纹、杯体光洁、厚薄均匀的郁金香花型的高脚玻璃杯。（专用品酒杯：杯体高度为54±2mm，最大直径为45±2mm，上口内径为33±2mm，满口容量为50-55ml。）酒杯应该充分清洗干净，并放置于不被其他气味污染的环境中。

品鉴者： 品鉴者在品酒之前应该注意清洁口腔，不吃重口味的食物以免对口感造成影响。品鉴者不宜抽烟、喷香水或使用香味浓的化妆品，以免影响对白酒香气的判断。品鉴者品鉴前夜不应熬夜，保证充足的睡眠，以保持感官的敏锐度。

其他物料的准备： A4白纸，以便观察酒色。矿泉水，以便清洗酒杯或者清洁口腔。白味面包或饼干，以便品鉴者多次品鉴时清晰口腔，恢复口腔敏感度。笔记本和笔，以便品鉴者及时做好品鉴笔记。还有无香味的纸巾、吐酒桶等。

白酒品鉴四部曲

当样酒上齐之后，首先注意酒杯中的酒样的多少，要使各酒杯中的酒样基本相同。白酒品评要遵循一定的顺序，即：先品鉴颜色浅的再品鉴颜色深的；先品鉴香气淡的再品鉴香气浓的；先品鉴酒精度数低的再品鉴酒精度数高的。确定好品鉴的顺序后，我们就可以依次对杯中酒进行品评了。品评一杯酒分为四个基本步骤：观色、闻香、品尝、评价。

观色

先把样酒放在评酒桌的白纸上，正视和俯视酒杯，观察酒样的色泽，色泽为无色或微黄。再观察透明度，透明度为清亮透明或者浑浊。再观察有无悬浮物和沉淀，把酒杯拿起来，然后轻轻摇动，观察酒液挂杯，悬浮物为有或无，沉淀为有或无。将色泽、透明度、悬浮物、沉淀的观察结果做好记录。

闻香

闻香是品酒中关键的一步。先将酒杯端在手中，鼻子与酒杯保持一定的距离（1-3cm），对酒样进行初闻。再用手扇风闻，然后将其接近鼻孔进一步细闻。在闻时一定要注意先呼气再对酒缓慢吸气，不能对酒呼气。每杯酒按上述步骤最多闻3遍，每次都要有准确的记录。闻完一杯酒后，稍事休息，再闻另一杯。若样酒较多，可先按1,2,3,4,5，再按5,4,3,2,1的顺序反复闻几次。先选出风格最明显的，再反复比较风格比较相近的，不断修正记录。

香气判断指标有：香气是否纯净，即是否有异味；香气的浓郁度，香气是浓郁还是淡雅；何种香型；香气的持久度。

品尝

品尝时，要注意将暴香或异香味酒样放置最后品评，以免干扰对其他样酒的品鉴。一般开始先含小口样酒（0.5-2ml），酒液布满舌面后，仔细辨别其味道，停留5-10s后，将酒液吐出或咽下。然后使酒气随呼吸从鼻孔呼出，判断酒气是否刺鼻以及感受香气的浓淡。并用舌头进一步品尝酒味是否协调。移动舌面与口腔上下摩擦，感受酒液中的涩味。再适当增加饮量，以判断酒液回味的长短，余味是否纯净，是回甜还是后苦，有无余香，以及是否有刺喉和不快之感，注意要边尝边做记录。品尝次数不宜过多，一般不超过3次。每次品尝后用水漱口，或吃些无味的面包或饼干，清洁口腔，防止味觉疲劳。

同种香型不同等级的酒，其口味的优劣主要表现为是否绵软、协调、醇厚、丰满、爽口、余香与回味等。任何香型的白酒，如出现后苦，酸涩和邪杂味感等，其品质则不高。

口感指标

柔和度：醇和，柔和，平顺，平和，辛辣，燥辣

丰满度：浓厚，丰满，醇厚，饱满，丰润，厚重，平淡，清淡，淡薄，寡淡

谐调度：谐调，平衡，细腻，粗糙，失衡

纯净度：爽净，净爽，涩口，欠净

持久度：绵长，长，短暂

酒精度：低：40度以下　　中：41-50度　　高：51度以上

评价

根据上述三个步骤的观色、闻香、品尝之后即可对酒做出综合判断。即酒的风格、酒的品质、酒的价格区间等。最后根据百分制对所品酒款做综合打分。

典型白酒风格

浓香型风格：浓香型白酒在窖池中发酵，是最知名、范围最广的白酒种类。可以是单粮也可以是多粮酿造。独特的地理环境和酿造技术让酒香独具一格，如以四川五粮液、四川泸州老窖以及四川水井坊酒为代表的产品具有香气浓郁、绵甜甘爽的特点，而在浓香型白酒中以五粮液最为出名。

酱香型风格：酱香型白酒的名称源自酒中别具一格的酱香。酱香突出、幽雅细腻、酒体醇厚、回味悠长、空杯留香持久。以贵州茅台酒、四川郎酒、湖南武陵酒等老品牌为代表，但现在市场上的时尚酱香新秀——四川五粮液股份公司出品的"永福酱酒"、四川舍得酒业有限公司出品的"吞之乎"也独具特色，具有酱香型白酒典型风格。

清香型风格：清香型白酒是用大麦和豌豆制成的大曲发酵，清香纯正、醇甜柔和，以山西汾酒和北京二锅头为代表。清香型白酒自然谐调、余味净爽，酒体具有清、爽、绵的风格特征。

米香型风格：米香型白酒以广西桂林三花酒为代表，蜜香清雅、入口柔绵、落口爽冽、回味怡畅。以长米或糯米或者两者混合为原料，与小曲进行发酵，最好的米香型白酒入口柔绵、醇和甘滑。

豉香型风格：豉香型白酒的名称源自中国一种用发酵的大豆制成的调味料——豆豉。该钟白酒以广东石湾酿酒厂的玉冰烧酒为代表，豉香独特、醇和甘滑、余味爽净。

凤香型风格：凤香型白酒以陕西省凤翔县的西凤酒为典型代表，它结合了浓香型白酒和清香型白酒的特点。西凤酒要经过其特殊容器——酒海储存。制作酒海时，酿酒师要用菜油、蜂蜡、血料等反复涂擦、晾干。西凤酒醇香秀雅、醇厚丰满、尾净悠长。

兼香型风格：兼香型白酒是通过多种工艺酿造或者两种或者多种不同类型白酒的混合。酱中带浓的风格以湖北白云边酒为代表。浓中带酱的风格以黑龙江玉泉酒为代表。

老白干香型风格：老白干香型白酒跟清香型白酒较为相似，但是不同于清香型白酒，它是以大麦和豌豆为原料制成的曲来发酵，陈年时间少于六个月，醇香清雅、甘润挺拔、丰满柔顺、口味悠长、风格典型，以衡水瑞天酒业生产的老白干白酒为代表。

芝麻香型风格：芝麻香白酒芝麻香突出、幽雅醇厚、甘爽谐调、尾净，以山东景芝酿酒厂的景芝白干和江苏梅兰春为代表。

特香型风格：特香型白酒酒香芬芳、酒味醇正、酒体柔和、诸味谐调、香味悠长。以江西"四特酒"为代表。

药香型风格：药香型也称为董香型，以贵州"董酒"为代表。药香型白酒药香舒适、香气典雅、酸甜味适中、香味谐调、尾净味长。

馥郁香型风格：馥郁香型白酒以湖南"酒鬼酒"为代表。酒体清亮透明、芳香秀雅、绵柔甘洌。

品评技巧

我们对酒的第一印象，往往是最正确的，人的味觉、嗅觉极易疲劳，所以品酒时间不宜太长。

品鉴白酒时，七分靠闻，三分靠尝。也就是70%以上可以从香气看出酒的品质。三七规律不表示香气大的酒就是好酒，香气的判断要以协调、优雅为基本标准。如果香气优雅细腻，自然纯正，酒的品质一般也不会差。

切忌闻香完成之前喝酒，更不能闻一杯喝一杯，一旦喝下一杯酒，后面基本是闻不出什么味的，因为味觉感受到刺激后会引起嗅觉的"失聪"。

要注意养成一边品鉴一般记录的习惯，俗话说好记性不如烂笔头，在品鉴过程中会去捕捉感官的体验，但品鉴时会感受到很多信息，很多感觉是一闪而过的，这些微妙的感受要随时记录下来，以免品鉴结束后不知所以。

品鉴笔记表

品鉴者：		地点：			日期：		
观色	颜色	异色		微黄色		无色	
	透明度	浑浊			清亮透明		
	悬浮物	有			无		
	沉淀	有			无		
闻香	纯净度	有异味			纯净自然		
	浓郁度	淡		中等	浓		浓郁
	香型特征	原料香		发酵香		陈年香	
	持久度	短暂		长		绵长	
品尝	酒体	单薄		中等		厚重	
	柔和度	燥辣	辛辣	平和	平顺	柔和	醇和
	丰满度	寡淡	清淡	丰润	饱满	醇厚	浓厚
	谐调度	失衡	粗糙	细腻	协调	平衡	谐调
	纯净度	欠净		涩口		净爽	
	持久度	短暂		长		绵长	
	酒精度	低：40度以下		中：41-50度		高：51度以上	
评价	风格	酱	浓 清 米	豉 凤	兼香 老白干	芝麻香 特香	药香型 馥郁香型
	品质	低劣	平庸	一般	中等	好	极佳
	价格	低端酒		中端酒		高端酒	
打分	百分制	60分以下	60-70	70-80	80-90	90-100	

Section B

1 B   2 D   3 A   4 B   5 C   6 C   7 D   8 D

2 A professional Baijiu taster is not born with a subtle sensation which, in most of time, derives from autonomic and systematic training. When reacting to authentic tasting, people will become

more sensitive toward different aroma, where abilities of distinguishing and memorizing various aromas are improved. Sensation can be trained by using daily materials such as honey, oil, soy sauce, flowers, rice, sesame, vinasse, etc. To be a qualified Baijiu taster, all one needs are incessant practices.

Unit 7

Section A

I.

- a Eight activities are highly valued in China: music, chess, calligraphy, painting, poetry, drinking, flower and tea.
- b Chinese people drink in three ways: drinking alone, drinking in pair and drinking in group.
- c The intention of the drunkard lies not on the liquor, but on other purposes.
- d Chinese people would like to express their moods. For example, misery.
- e "Surrounded by distinguished guests, immersed in fragrance of osmanthus in floor vases, we indulge ourselves in drinking all night long."
- f The host will urge his guests to drink.
- g Chinese people honor nature and the ancestors. Not only is Qingming a period for commemorating the dead. It is also time for people to go out and enjoy nature.
- h calamus liquor, realgar liquor, Toad liquor, albizia liquor

III.

a—3 b—5 c—9 d—6 e—2 f—10 g—7 h—4 i—8 j—1

IV.

1 commemorate
2 recklessly
3 withered
4 parallels
5 ritual

6 spectacle
7 ominous
8 hilarity
9 suicide
10 indulge

V.

中国人好客，在酒席上发挥得淋漓尽致。人与人的感情交流往往在敬酒时得到升华。人们在敬酒时，都想请对方多喝点酒，以表示自己尽到主人之谊。客人喝得越多，主人就越高兴，说明客人看得起自己；如果客人不喝酒，主人就会觉得有失面子。

文敬：是传统酒德的一种体现，即有礼有节地劝客人饮酒。

回敬：这是客人向主人敬酒。

互敬：这是客人与客人之间的敬酒。为了使对方多饮酒，敬酒者会找出种种必须喝酒的理由，被敬酒者若无法找出反驳的理由，就得喝酒。

译文

中国人八大雅事：琴棋书画诗酒花茶。饮酒是一种礼仪，一种文化，一种入世的态度，出世的智慧。中国人饮酒大体上有三种形式：独酌，对饮，群饮。中国人重心境，酒只是抒情达意的工具而已。喝酒也是"醉翁之意不在酒"。独酌既有李白"花间一壶酒，独酌无相亲"的雅味；又有韩奕"相会年年少旧人，独酌杯中酒"的伤怀。酒逢知己千杯少，话不投机半句多。白居易渴望与朋友把酒共饮，互诉衷肠，才有了"绿蚁新醅酒，红泥小火炉。晚来天欲雪，能饮一杯无？"苏轼更是真率豁达地同友人对饮，"醉笑陪公三万场，不用诉离殇"。中国人喝酒讲究的是呼朋引伴、开怀畅饮。晏殊用"座有嘉宾尊有桂，莫辞终夕醉"来描述酒局里一大群人的狂欢。赏花饮酒，李白和友人快意人生，"落花踏尽游何处，笑入胡姬酒肆中"。

《红楼梦》更是以酒为契机，生动细致地描绘了中国酒文化。如第一回里甄士隐和贾雨村的豪饮："二人归坐，先是款酌慢饮，渐次谈至兴浓，不觉飞觥献斝起来……二人愈添豪兴，酒到杯干"。劝酒也是中国酒文化的一部分。中国人的酒桌上要不停地劝酒，才能显出热情，才能有饮酒的气氛。如第44回王熙凤过生日时鸳鸯劝酒。"真个的，我们是没脸的了？就是我们在太太跟前，太太还赏个脸儿呢。往常倒有些体面，今儿当着这些人，倒拿起主子的款儿来了。我原不该来．不喝，我们就走。"鸳鸯劝酒，展现了她的热情，加深了她们之间的友谊。宴会劝酒，宾主尽欢。中国人在喝酒时，尤其是寒冷的天气喝酒时，会把酒温热一下再喝。如第七回薛姨妈劝宝玉不要喝冷酒。"难道就不知道酒性最热，要热吃下去，发散的就快；要冷吃下去，便凝结在内。拿五脏六腑去暖他，岂不受害？"

美酒助兴，一生如此，一年亦是如此。酒让生命的过程贯穿在春夏秋冬，以四时节序的严格遵守让其始终保持柔雅的"中国味道"。如除夕夜的"年酒"。过年，也叫除夕，是农历年的最后一天，人们常常熬夜或通宵不睡，畅谈过往，展望未来。这是家人团聚的日子，年夜饭是一年中最为丰盛的酒席。一起吃吃喝喝，一家人其乐融融。阳历4月5日的清明节，也称为扫墓节，中国人会祭祖、踏青。 此习俗可追溯到古代，现今仍很重要。清扫墓地，用死者生前最爱的食物，酒，筷子和纸钱来祭奠死者是对死者表示敬意的两项惯例，希望死者来世生活幸福。清明节不仅是祭奠死者的日子，也是人们外出郊游的好时光。人们借酒和自然美景来平缓或暂时麻醉人们哀悼亲人的心情。农历五月五日的端午节，是人们纪念古代自杀的爱国诗人屈原的传统节日。人们为了辟邪、除恶、解毒，有饮菖蒲酒、雄黄酒的习俗。同时还有为了壮阳增寿而饮蟾蜍酒，镇静安眠而饮夜合欢花酒的习俗。农历八月十五的中秋节，更是家人朋友赏月饮酒的日子。中国人相信，满月象征着家人团聚。此时人们偏爱桂花酒。《天宝遗事》记载，唐玄宗在宫中举行中秋夜宴，灯烛熄灭，音乐弥漫。他们进行"月饮"。农历九月九日重阳节，人们登高、赏菊、饮菊花酒。九月秋高气爽，适合郊游。浸入茱萸果的菊花酒，更有镇痛功效，让人元气满满。这些习俗至今不衰。

Section B

Choose the best answer to complete each of the following statements.
1 B　2 C　3 B　4 D　5 C　6 A　7 A　8 C

  Liquor plays an important role in functions such as wedding ceremonies, funerals, birthday parties, worshipping rituals, entertaining activities. It seems that a liquor-free worshipping ritual, a commemoration ceremony or a praying session for good harvests would have no carrier for people's sentiments; a liquor-free wedding ceremony would have no vehicle for the bride and the groom to vow for their mutual love; similarly, a liquor-free funeral would make the descendants feel nowhere to demonstrate their grievance for their beloved ancestors; a liquor-free birthday party would be difficult to show people's etiquettes and interests; a liquor-free farewell party for soldiers going off to the battlefield would be difficult to express the heroism and the tragic feelings for those leaving. In short, customs would not establish themselves without liquor. liquor has served as an important base and support to the formulation of folk customs.

译文

中国是个多民族国家，少数民族的酒文化也是异彩纷呈。

蒙古人能歌善舞尚酒。在举行各种隆重典礼时，先要由地方首领或德高望重者端起酒碗，用右手无名指蘸上酒，朝天弹一下，意为敬天神；再蘸上酒朝地弹一下，意为敬地神；再蘸上酒，面朝西方弹一下，意为敬祖先。在蒙古人心目中，马奶酒是神圣的饮料。招待客人什么东西都可以缺，但绝对不可以缺酒。以酒待客，是蒙古族最普遍也是最起码的礼仪。主人拿出酒杯或者银碗，给客人斟酒。主人将盛酒的碗托举在长长的哈达上，给客人连献三碗。第一碗感谢上苍恩赐我们光明，第二碗感谢大地赋予我们福禄，第三碗祝人间永远吉祥。常见的情景是，主人恭恭敬敬地把酒双手捧给你，此情此景让你无法不喝。若你脖子一仰痛快喝下了，主人接着给你敬第二碗，你肯定找理由推辞。遇见这种情况，主人会朝你笑笑，而后再给你唱支歌，边唱边把酒杯举过头顶。人家如此敬重你，不能不接酒杯；若你接过来了就不能放下，必须喝下，这是规矩。敬第三碗时，主人先是唱着祝酒歌双手举着酒单腿跪下，你若不接酒，主人就双腿跪下，盛情难却。客人连饮三杯后，斟上第四杯时，客人便一手端起酒杯一手高高扬起，或向主人说上几句祝福的话，或放开嗓门唱起祝福主人的赞歌。蒙古族人以"醉客为敬"，客醉证明主人的敬客之意已到，客人不醉不散。客人喝醉，主人认为是看重他，非常高兴；客人对敬酒推让不喝，会被认为是瞧不起主人。在蒙古人眼里，酒是饮料的精华，敬酒是对客人的欢迎和尊敬。蒙古人敬酒，常不备菜，而以美妙动听的酒歌劝酒。歌手会根据不同场合，选择适合的歌词，酒歌悠扬清亮。

彝族人十分好客，也嗜酒，男女老少皆能饮酒。彝族谚语曰："汉人贵在茶，彝人贵在酒"。"为人不喝酒，白来世上走。""有酒便是宴。"因此常常只喝酒，不一定有菜。彝族人喝"转转酒"更有特色。他们不分生人熟人，不分地点场合，更不需要菜肴。大家席地而坐，围成圆圈，把酒倒进大碗里，一边谈天，一边依次轮流饮酒。你喝一口递给我，我啜一口传给他，大家依次轮流喝着这一碗酒。酒在彝族生活中是不可或缺的，它是一种文化，一种习俗和一种标志。酒除了是生活习俗外，还常用以表示决心和信誓。为了表示对某件事情干或不干进行发誓时，他们常用饮血酒来表示。即当场杀一只大公鸡，把鸡血滴进酒中，双方或几方当事人端起酒碗，进行发誓。誓毕，则一饮而尽，表示永不反悔。在节假日，随处可见彝族人从家里携带咂酒坛，坛中插一根或数根咂管，男女老幼，三五成群聚做一堆。边咂边谈，纵谈天下事，细数人中情，谈笑风生，情助酒性，酒助情深。这坛酒有化陌为友，增亲朋为莫逆和成就婚姻的功效。由于咂酒有着强烈的民族气息，一直长盛不衰。

广西田东县的壮族人，还有一个"羹来羹去"的饮酒习俗，这是世界上"独一无二"的饮酒方法。他们不习惯直接使用大碗喝酒，也不喜欢使用酒杯，他们常用的是匙羹。喝前先把酒倒到一个大碗里，然后主客互敬。主人用一个匙羹，从大碗里打一羹酒，送到客人嘴边，对方是必须要喝下的，不然就是不给"面子"。喝了别人敬的酒，就要回敬。客人也拿起另一个匙羹，从大碗里打一羹酒，"灌"进主人的嘴里。不管喝多长时间，都使用这样"你来我往"的方式，这就是所谓的"羹去羹回"。

侗族是一个热情好客的民族。和其他民族一样，少不了以酒为礼，以酒待客。侗族人的拦路酒是主人们或在寨门，或在街巷，或在道口，或在鼓楼下，或在桥头，或在家门前摆上路障，拦住客人的去路。在路障后面，是盛装的侗族小伙子和姑娘们。他们一个挨着一个横排站着，客人必须和姑娘们对歌，饮下一大碗家酿的米酒。如果不唱歌，不喝酒，那是休想过去的。

Unit 8

Section A

I.

a In China, there are three major categories of Baijiu, including sauce aroma, strong aroma, and light aroma.

b Sauce-aroma style Baijiu is a highly fragrant distilled sorghum liquor of bold character and provides a flavor resembling Chinese fermented bean pastes and soy sauces.

c From natural conditions in the town of *Moutai*, *Moutai* has the unique and superior advantages, particularly climate and vegetation that are beneficial to make the liquor. *Moutai* is distilled from fermented sorghum that is mainly picked out of glutinous sorghum, commonly known as *Red Tassel Sorghum*.

d Due to its unique characteristics, *Moutai* gained many awards and prizes at home and abroad, particularly a gold medal at the 1915 Panama-Pacific Exposition in San Francisco, California. Thus it has gradually been a kind of "first impression" to make westerners get better understanding of China. Usually, it is one of Baijiu served in China's official state banquets to foreign heads of state and distinguished guests visiting China, and even it is the only liquor presented as an official gift by Chinese embassies in foreign countries and regions. More importantly, *Moutai* has been considered as a diplomacy to maintain and express the image of China.

e National Cellar 1573 is the perfect example to present the highlighted features of Luzhou Laojiao.

f Literally, Wuliangye refers to five grains. Exactly, it contains the five main ingredients, including sorghum, corn, glutinous rice, long-grain rice, and wheat.

g It has a unique and comprehensive style of long-lasting fragrance, mellow and rich taste, clean and pure flavor entering the throat, and balanced flavors in harmony, for it owns a number of materials.

h In the Chinese history, Fenjiu totally experienced three glorious periods. Firstly, it could be dated back to the Northern and Southern Dynasties (550 A.D.) and was recorded into the *Twenty-Four Histories* for gaining the popularity of the Emperor of Northern Qi Dynasty. Secondly, Xinghuacun was featured in the poem "Qing Ming" written by the great poet Du Mu (803-852 A.D.) in the late Tang Dynasty, which is now a part of mandatory reading for primary school. Exactly, due to the famous poem, Fenjiu gained great fame, also called as "Xinghuacun Liquor". Finally, in 1915, Fenjiu won a gold medal at the 1915 Panama-Pacific International Exposition, representing a milestone in the development of Chinese Baijiu and a leader of Chinese Brewing Industry.

III.

1 j   2 g   3 e   4 i   5 d   6 h   7 b   8 c   9 a   10 f

IV.

1 resembles   2 feature   3 classified   4 conventional
5 superior   6 entertain   7 renowned   8 unique
9 beneficial   10 consolidated

V.

作为中国白酒的优质核心区，贵州省拥有两大全国知名品牌，分别是茅台与董酒。在这两大品牌之中，董酒得名于1942年，并在白酒市场享有盛名。在蒸馏酒行业中，董酒从原材料到加工技艺已逐渐形成了自身的风格特点，特别是将大曲与小曲进行了完美的融合。董酒因传承着中国古代文化的深厚根基而被视为民族瑰宝，还看作是中医传统几千年来真正的活化石，并成为中国最显著的标示符号。

20世纪八十年代，董酒在优质白酒的较量中，堪比茅台、五粮液、泸州老窖。因此，自1963以来，它在以品评中国白酒质量为目标的众多年度大会上，赢得了许多奖项和金质奖章，甚至在海外获得了世界级的荣誉。并且，因其独特的风格与原创性的科学配方，董酒被国家权威部门永久列为"国家机密"。在2008年，董酒被正式认定为当地标准，以及"药香型"白酒的典型代表。

译文

中国名酒选录

白酒以独特的香味和口感而闻名于世，是中国最具地方性的特色之一。因其丰富多样的香型特征，大众可以对白酒进行鉴别并对其加以细致的品评。更为准确地说，白酒已被更科学地分类为不同的香型，并经由中国不同地区多样化的生产工艺，进而满足各类消费人群的需求。如今，白酒主要有三大类，即酱香型、浓香型和淡香型。显然，这三种酒的香气特征比一些小众风格的白酒更具有典型性和代表性，并且其比重约占中国白酒市场的60%-70%。

这三种类型皆有着知名的品牌来充分展现自己；反过来，他们的品牌又对其自身产生了长期有益的影响。普遍认为，品牌可以促进白酒知识不断丰富。因此，了解那些著名的白酒品牌对于公众而言是极为重要的。本文从市场上所有品牌中挑选出不同类型的知名白酒，展示它们各自的特征。

1. 酱香型

第一类是以酱为基础的香型，这是一种高度芳香的蒸馏高粱酒，具有鲜明的特征。并且，其提供了一种类似于中国发酵豆瓣酱和酱油的味道。事实上，因最著名的白酒茅台，其曾被称为茅香。这就是为什么茅台被认为是酱香型白酒的鼻祖。因此，当谈及酱香型白酒时，茅台就是其最完美的示例。

茅台或茅台酒产于中国西南部贵州省仁怀市附近的茅台镇，以其独特的酱香特质而成为世界知名白酒品牌。茅台与苏格兰威士忌、法国科涅克白兰地齐名，是世界三大蒸馏酒名酒之一。并且，茅台酒也因其悠久的历史而被盛誉为中国的国酒。

从茅台镇的自然条件来看，当地具有独特的酿酒优势，尤其是当地的气候和植被。这些较好的条件仅仅为其酿造打下基础。茅台酒具有独特而系统化的酿造工艺，主要从俗称红穗高粱的黏性高粱中提取而得，再经由发酵高粱而蒸馏所成。在一系列的酿造过程中，蒸馏是极为重要的。这意味着由曲和高粱所组成的发酵混合物在一年内经历蒸馏七次，且每批都储存在一个单独的容器中。但是，白酒的香气会随着蒸馏季节的不同而略有变化。茅台酒现存不同的版本，酒精含量也从53%到35%不等。

茅台酒已有200多年的酿造历史。在清朝（1644—1912年），中国北方的酿酒师将先进的技术引进到当地的酿造工艺中，从而创造出了一种独特的白酒，这就是茅台的起源。此后，茅台酒在当地几家酿酒厂进行生产。此外，在中国内战期间，人民解放军选择在茅台镇扎营，并采取了一些措施来管理当地的白酒。由于中国共产党取得了内战中的胜利，政府将当地的所有酿酒厂合并为一家综合性公司。最终，一家具有部分上市的、部分国有的企业，贵州茅台由此成立。并且在1951年，中华人民共和国成立两年后，它被誉为"国酒"。渐渐地，它已成为更受人们欢迎的选择。

1915年在加州旧金山举行的巴拿马太平洋博览会上，茅台酒荣获一枚金质奖章，从此闻名于世。与此同时，它还在1985年和1986年的巴黎国际博览会上分别获得两次金奖。国内外不断增加的曝光率，使得茅台酒成为西方人更好地了解中国的第一选择。并且，它是中国宴请外国国家元首和贵宾的国宴上常用的白酒之一，甚至是中国驻外使领馆作为官方礼物赠送的唯一一种酒。例如，在1972年美国总统访华国宴期间，周恩来总理用茅台招待理查德·尼克松。周总理告诉尼克松，茅台酒自1915年赢得认可以来就变得很有名了，并且在长征中"茅台酒被我们用来治疗各种疾病和伤口"。尼克松回答说："让我用这灵丹妙药敬酒吧。"茅台酒被誉为"外交之饮"，在中国政界和商界领导人的公务接待或其他活动中，其一直是最受欢迎的白酒品牌。

此外，茅台酒目前销往全球100多个国家和地区，总销量超过200吨。自2017年4月以来，该公司超越帝亚吉欧，成为全球市值最高的酒类公司。因此，贵州茅台也正在制定全球扩张计划。

2. 浓香型

浓香是第二种与其他品种的白酒表现不同的香型。它实际上是指一种味道甜美、质地醇厚柔滑的蒸馏酒，这种酒温和持久的芳香来源于高水平的酯类，主要是乙酸乙酯，进而带有较为浓郁的菠萝、香蕉和茴香等味道。与酱香型不同，浓香型具有一个明显的特征，即大多数白酒皆从高粱中蒸馏而出，且可与其他谷物混合蒸馏，并融合在窖池中持续发酵。事实上，浓香型还有另一个名字，"泸香"，因其由泸州市泸州老窖酿酒厂所独创而得。因此，泸州老窖可以完全展现其特色。此外，宜宾的五粮液，绵竹的剑南春和宿迁的洋河也是这类白酒的典型代表。

泸州老窖是中国最受欢迎的白酒之一，其历史长达400多年。其酿酒厂甚至可以追溯至秦汉时期，而兴盛于明朝1573年。泸州老窖起源于古代一批最古老的白酒作坊，显然已成为中国现存最古老的白酒之一。据大量完整的史料记载，泸州老窖常被视为中国白酒的发源地，该酒现由总部位于中国四川省泸州的大型国有企业泸州老窖有限公司独家生产。泸州市也因此赢得了很高的声誉，被誉为"白酒之城"。

泸州老窖作为浓香型白酒的典型代表，最早以"浓香"为基础酿制而成，并具有优良的传统酿造工艺，于2006年被列为国家级非物质文化遗产。此外，泸州老窖还有一个非常著名的标志，能让人从众多白酒中轻易辨认而出，即国窖1573。国窖1573起源于1573年建立的一组酿酒窖池。迄今为止，泸州老窖有着固定的酒精含量，基本在42%到52%之间。

得益于气候和水源等优越条件，泸州老窖以其带有独特香气和口感的蒸馏品质而闻名。具体而言，它是在具有独特黏土成分的环境中酿造的，这样的黏土赋予了其著名的香气和调色板。泸州气候温和，极端温度在40.3℃~-1.1℃之间，年降雨量丰富，年平均风速稳定。这些气候条件使其成为培育地域性独特的农作物品质及微生物类群的最佳场所。换言之，泸州本地糯红高粱能够通过这种独特的气候得到更好的滋养，并以此为泸州老窖精心挑选出其所需原料。然而，除了适当的气候，水源是酿造更好白酒的另一部分。通过一系列科学分析，来自长江的酿造用水具有明显的无臭、微甜、弱酸性、适宜度数、充足微量元素等特点。确切地说，这些特性可以有效地促进酵母的繁殖，有利于糖化和发酵。特别是在大曲酒的酿造生产中，酶促反应可以大大得到提高。

泸州老窖具有浓郁的蜜桃发酵香气（酒精含量超过50%），是国内外最好的白酒之一，甚至是中国最古老的四大名酒之一。对于泸州老窖来说，其标志性产品国窖1573的畅销，一直在推动着整体的销售增长。该公司表示，2018年总销售收入可能同比增长25%以上，主要受到高端产品销售增长的推动。因此，泸州老窖更注重其高品质作为增加销售的一种方式。正如其网站所言："我们的理念是保持生产的完美，尽我们最大的努力让消费者喝到高品质的白酒。"

浓香型白酒的另一显著代表是五粮液，它产自四川南部宜宾市，是一种具有浓郁香气的陈年蒸馏酒，其酿造用水主要来自岷江中部。关于它的历史渊源，可以追溯至南北朝时期（公元420年-589年）。从汉字的角度来看，"五粮液"字面上是指五种谷物，而其中恰好含有高粱、玉米、糯米、长粒米、小麦等五种主要成分。选择五种谷物作为原料，该配方至少在明代就已经问世，且于1905年因此得名五粮液。保留下来的最古老的酒窖已经超过600年，至今仍在使用。五粮液逐渐走向国产化、标准化，平均酒精含量从35%上升到68%。

五粮液采用传统酿造工艺，选用五种精选谷物进行发酵。它因使用了多种材料，而具有香气悠久、口味醇厚、入喉净爽、各味谐调的独特而全面的风格。五粮液以其独特的生态条件，赢得了良好的声誉，并越来越受到国内外市场的重视。此外，它强调保持传统是生产优质白酒的基础，被誉为目前各类白酒的杰出珍品。

由于这些优势，五粮液多次获得"国酒"金奖。1991年被评为中国"十大驰名商标"。然后，在1915年巴拿马太平洋博览会上荣获金奖80年后，五粮液分别再次在1995年和2002年的国际博览会上获得唯一的金牌。由于1995年获得奖章，它被评为"中国酒业大王"称号，并奠定了其在中国白酒行业中的辉煌与成就。到目前为止，五粮液已在世界范围内的各种展览或比赛中荣获了无数的荣誉和奖项，并以其独特的酿造工艺，不断地书写着它传奇的历史。

作为优质品牌，五粮液正日益走向全球化，其成长已成为一个"脱颖而出"的故事。根据伦敦市场研究公司在《中国日报》发布的一份在线报告显示，五粮液是2018年增长最快的品牌，也是白酒行业中的冠军，其排名上升184位直至第100位，已达品牌价值146亿美元，与同比增长161%。此外，宜宾五粮液股份有限公司表示，在2018年第一季度，其各类产品的白酒销售量为6000吨，高于过去几年每季度的销售量。五粮液集团有限公司已被公认为世界优质品牌，致力于打造以白酒为中心的现代机械制造、现代材料科学、现代包装、现代物流等多元化发展的产业创新格局。

3. 清香型

作为第三种香气类型，清香型具有许多突出的特性：细腻，干燥，温和，口感醇厚、清爽。它甚至能给人以醇厚的甜味和清爽的回味。这类蒸馏酒的风味主要来自乙酸乙酯和乳酸乙酯之间的平衡，并使其具有干果和花香的味道。这种酒是从在石器中发酵的高粱与并用麦麸或大麦和豌豆混合制成的曲共同制成的。北京的二锅头和山西的汾酒是完美呈现这类白酒的两种典型代表，前者在台湾被称为高粱。然而，因在杏花村汾酒酿酒厂生产的清香型白酒被称为"汾香"，后者实际上才是清香型白酒的最好例子。

汾酒是原产于山西省汾阳市杏花村的蒸馏高粱白酒，已有近4000年的悠久历史，是中国最早的高粱白酒，在中国历史上有三个辉煌的时期。它最早可追溯到南北朝时期（公元550年），由于受到北齐皇帝青睐而被记录于《二十四史》。后来，杏花村出现在唐代晚期伟大诗人杜牧（公元803年-852AD）所写的《清明》诗中，这首诗也成为现在必读诗篇的一部分。正是因为这首著名的诗，汾酒获得了巨大的声誉，也被称为"杏花村酒"。最终，在1915年的巴拿马太平洋国际博览会上，汾酒获得了金奖，成为中国白酒发展史上的一个里程碑，以及中国酿酒业的领导者。与此同时，汾酒历史悠久，与华夏文明、黄河文化、晋商文化同根同源，使汾酒能够在各种白酒中发展出自己的香气特征，并成为白酒界公认的最具竞争力的优势之一。作为国宝，汾酒凝聚了古代中国人民的智慧。

在品种繁多的白酒中，山西汾酒是清香型白酒的鼻祖，甚至成为国家清香型白酒标准制定的最佳典范。它也是当地最受欢迎的酒，通常比其他中国北方酒更甜。作为清香型白酒的典型代表，汾酒有着深厚的传统和高超的工艺，使人人口即能享受到香甜的口感，饮用后能产生长久的余味。更为准确地说，通过从山西省中部精选本地高品质的高粱、大麦、豌豆作为原料，大多数熟练的酿造大师通常会创造出一种由大麦和豌豆制成的糖化发酵剂，并采用"清蒸二次清"的独特酿造工艺。由此可见，汾酒在酒精含量在38%至65%之间所呈现的独特而精确特征，是口味纯正温和，甘甜醇厚，回味爽口。

此外，那些技艺精湛的工匠们，从制曲到发酵，再到蒸馏，对于整个过程都有着其自身独到的见解。数百年来，这些技能通过师徒关系的口头交流而代代相传，并不断得到创新与发展。

另一个因素是在酿造汾酒时起至关重要作用的水源。水是白酒的主要成分，曲是白酒的主要结构。俗话说"名酒产地，必有佳泉"。从现代科学研究中可以发现，汾酒的秘密来自丰富而优质的地下水资源，其中含有多种有益于人体健康的化学元素。因此，清代著名诗人、书法家、中医学家傅山先生曾题词高度称赞汾酒的健体疗效。因此，汾酒以其优良的传统工艺和现代科学技术以及优质水资源而闻名。

目前，汾酒不仅是山西白酒产业的文化象征，也是中国白酒的文化象征。汾酒在1500年前已然成名，且在政府的扶持下逐渐形成了自己的白酒产区。随着山西经济和旅游业的发展，汾酒建成了一个文化景区，其特色为古建筑群落，总占地面积158万平方米，相当于10座故宫博物院。100多座雄伟的黑瓦建筑毗邻，并以唐、宋、明、清等朝代建筑元素为基础建造而成。在博物馆里，游客可以品尝样酒，并参观从酿造到储存、灌装和包装的整个生产过程。汾酒文化景区以及被认定为国家级非物质文化遗产的汾酒酿造工艺，支持了当地汾酒产业和工业旅游业的发展。

Section B

I.
1 B   2 F   3 G   4 F   5 C   6 A   7 D   8 E   9 B   10 D

II.

Jiannanchun is one of the typical representatives of Strong-aroma Baijiu in China, produced in Mianzhu city, Sichuan Province. Usually known as "The hometown of liquor", Mianzhu is famous for producing liquors and is also the birthplace of Sichuan Baijiu. Due to its distinctive and superior natural conditions, Mianzhu provides a better environment for brewing Jiannanchun. In addition, there has a long history and rich culture of Baijiu. According to historical materials and cultural relics of Mianzhu, it could be seen that Jiannanchun is not only a significant part of the history of Sichuan Baijiu but also a preciously cultural heritage of China.

译文

小众风格白酒

众所周知，中国白酒种类繁多，尤其以三种香型风格为主。根据经济学家和研究人员近来的研究证实，这三种香型极具代表性，并且在白酒市场中占有较大份额。这三种香型包含酱香型、浓香型、清香型三种风格，其生产工艺已实现标准化和固定化。除了这三类香型以外，采用不同的工艺技巧可以生产其他特定的香味和香型，比如三花酒，西凤酒和白云边。这些白酒都不属于前三类，而具有自身的香型风格，具体如下所述。

1. 米香型

作为中国四大传统白酒之一，米香型白酒具有蜜香轻柔，优雅纯净，回味怡然等特点。因此也称为"蜜香型"。这种白酒的高级醇含量较高，其乳酸乙酯含量大于乙酸乙酯含量，决定了其主体香。这种白酒精选优质大米为原料制作小曲进行发酵，米香型白酒采用半固态发酵为其主要生产工艺，未添加任何食用酒精及非白酒发酵产生的其他物质。这类由大米进行蒸馏所得的白酒，其主要代表为广西桂林所产的三花酒。据说，三花酒在广西桂林已有上千年的历史，且以添加天然芳香的草本植物和选用当地象山的清澈泉水而著称。其酒精含量为55%-57%。

根据最新标准，米香型白酒分为两类，即传统型和新型。传统的米香型白酒以桂林三花酒为代表，而新时代的米香型白酒则是以冰峪庄园大米原浆酒为特定代表。事实上，这种划分可视为是经济发展的需求和科学创新的产物。新时代米香型白酒不仅保持了米香型白酒的传统，而且还具有符合现代消费对各类健康影响的要求，诸如营养、食疗优质等。然而，值得注意的是，米酒与米香型白酒是截然不同的。前者主要是以家酿米酒的方式，在常温下一般具有一定的沉淀物；而后者则在工厂或是车间，采用消除所有沉淀物质的生产处理技术，用肉眼即可识别二者。

2. 凤香型

凤香型是指一种在窖池中发酵，置于藤制容器中陈酿而成的蒸馏酒。这种香型的白酒，皆具有类似于浓香型白酒的甘甜挺爽和醇厚悠长的特点。总体而言，凤香型白酒主要以高粱、大麦和豌豆为其原料，采用中温大曲或麸曲发酵，添加酵母混合制成糖化发酵剂。此香型白酒采用续馇配料法，进行土窖发酵，且窖龄不超过一年。凤香型白酒主体香味成分主要来自乙酸乙酯、己酸乙酯和异戊醇之间的平衡，其酒质特点为无色、清澈透明、醇香秀雅，甘润挺爽，诸味谐调。凤香型风格白酒以陕西省凤翔县的西凤酒为典型代表，其特点为入口甘甜、回味醇雅。

西凤酒因其历史悠久而成为中国最古老的名酒之一。它始于殷商，盛于唐宋，距今已有三千多年的历史。据中国的民间传说，凤翔县乃是凤凰的故乡，西凤酒遂因此而得名，寓意着白酒产于凤凰乡之西。在白酒行业的官方评估中，西凤酒荣获四次"国酒"称号。近年来，西凤酒十分重视革新和产品的研发，并在生产工艺上取得了较大的突破。基于自身传统工艺与香型，西凤酒致力于开发具有独特品味和不同酒精含量的系列型酒产品，以此迎合不同地区消费者的不同需求。与此同时，通过对其微量成分进行详细分析，可以证实西凤酒既不属于清香型白酒，也不属于浓香型白酒，而是具有独一无二的香型特点。

3. 兼香型

兼香型白酒又称为复香型、混合型，是指一种具有两种或多种类型白酒混合而成，且受到自然条件、生产原料和酿造工艺影响的蒸馏酒。这种白酒通常可以进行随机组合至少两种白酒香型，比如说酱香型、浓香型、清香型，进而有机地形成一种全新的类型。因此，兼香型白酒具有介于酱香型与浓香型之间的感官特征。这类复香型、混合型白酒的典型香味，来自庚酸、庚酸乙酯、乙酸异戊酯、甲基己基甲酮、异丁酸和丁酸等成分。为了满足现代人对科学膳食和口感的需求，兼香型白酒致力于改善酱香型白酒的粗糙后味，克服浓香型白酒香浓、口味重的通病。此外，它近来与其他两大香型白酒——酱香型和浓香型，成为深受消费者喜爱的白酒香型。

白云边酒堪称兼香型白酒的最佳代表，它产自中国中部地区的湖北省松滋市，其字面意思是指"白云酒的旁边"。白云边酒厂创立于1952年，已成为中国著名的白酒品牌，更是在过去60年里成为以品质赢得市场的典范。并且它采用中国传统的生产工艺，以至于现拥有7000吨酒的储存量。白云边酒的主要特点是一种浓酱兼香型的风格。实际上，作为兼香型的典型代表，它的香型特点接近于茅台酒。据2016年中国日报最新的一篇报道显示，白云边酒的销售收入达4.35亿元（约合6.53亿美元），且其税收配额为7.90亿元。因此，它深受全国人民群众的高度赞扬与偏爱，而非仅仅是在湖北省。